HISTORIC TRAILS ALBERTA

Mark Anderako

Lone Pine Publishing

First Printed 1985 5 4 3 2 1

Printed in Canada

The Publishers:
Lone Pine Publishing
414, 10357 109 Street
Edmonton, Alberta
T5J 1N3

Typesetting by Pièce de Résistance Typographers, Edmonton

Printed by Hignell Printing Ltd., Winnipeg

Canadian Cataloguing in Publication Data

Anderako, Mark
Historic trails Alberta

Includes index.
ISBN 0-919433-38-3

1. Trails —Alberta—Guide-books. 2. Alberta—Description and travels—1981- —Guide-books. I. Title.

FC3657.A64 1985 917.123 C85-090116-2
F1076.A64 1985

Original front cover watercolour by Elsie Baer

Maps: David Marko and Sheryl Caldwell

Archival photos courtesy Provincial Archives of Alberta

A narrow path, like a cow path, capable of receiving horses in single file. It is only 12 or 14 inches wide on dry ground and may be merely a smoothing of the ground, or it may be worn ankle deep by man and beast. It may be a mile, or a thousand, in length. It is apparently aimless. It seems to go nowhere, except the most difficult places; yet it is by no means lawless; it has designs — it ultimately gets there. It accomplishes wonders.

— Hamlin Garland

Contents

Contents

List of Maps

Acknowledgements

Thanks to: Grant Kennedy for giving me the opportunity to write this my first book. Mark Giles for his tireless editing. Margaret Barry for her preliminary research. Alberta Historic Sites for providing photographs, maps and research material. James MacGregor for his inspiring books. The trails enthusiasts who helped me find my way.

Introduction

Long before white men embraced Canada, Indians traversed its waterways. Most of their trails in the northern reaches of Western Canada were portages, linking a network of rivers and streams. And yet, though nothing influenced the patterns of exploration and settlement of Canada more than its waterways, trails were the conduits of a civilized scheme.

In the southern region of Western Canada, where rivers are less navigable, people of bronze wandered overland, generally along north-south trails, in their nomadic quest for food and shelter. They moved with the change of season and the tide of the herds. Though more than thirty thousand years had passed since their stone-age ancestors had migrated from Siberia across the glacial expanse of the last great ice age, it wasn't until the eighteenth century that the Plains Indians moved from travel on foot with dog-powered transport to horse-mounted travel. Now they could rule the plains.

These people revered the land, gleaned their living from nature's larder and altered their estate as little as possible. The early Indians did not blaze trails as did the interloping white men. Their trails were dimly marked, barely perceived by the uninitiated eye, silently running through parkland grasses, timbered lowlands and scrub thickets.

DOUG BABCOCK–ALBERTA CULTURE

The Athabasca Landing Trail (see pages 73 to 83) as it appears today, two miles south of Stony Creek near Tawatinaw.

New arrivals in a strange land, the Europeans adopted these trails and employed the Indians to guide them in their exploration and trading ventures. While the fur era boomed, competition pressed traders further west in search of new markets and the patronage of tribes camped far from the posts at Hudson Bay. Newly built supply and base forts — bastions of safety — lined the rivers. Trade, settlement and exploration followed these well-worn overland highways. Only later, when order was forced on the unruly west, did the pioneers dare cut more direct paths through the Indian's land.

Over time the fur trade ebbed, and the new Dominion of Canada began to place its stamp on the west. Canada was building a nation and homesteaders, prospectors, missionaries and whiskey merchants added to the maze of deeply-etched trails. Trails were broadened to accommodate travel by pack horse and cart. Alberta's soil was well advertised and transcontinental overland roads ushered in a swarm of migrating settlers. Soon the province was freckled with towns and rural hamlets. Many age-old trails were surveyed, improved and eventually paved with asphalt and girded with steel in preparation for high-speed mechanized travel. The trails that survived are rare and valuable.

Given that trails were the forerunners of our modern roadways, there is always a certain nostalgia invoked when remembering the simpler but dramatic events of history. Though a select few study our past in detail, the lure of historic trails is hardly confined to the academic realm. More so today than ever, we seek full value for our leisure time. Trails offer a host of recreational possibilities — rest, exercise, fast-paced thrills, or a refreshing stroll in pleasant — surroundings. In every way trails are an integral part of the natural experience. When fully developed with interpretive facilities, historic trails educate as they entertain.

Though it is only recently that Canada's trails have become anything more than transportation routes, other countries throughout the world have developed a respect and enjoyment of trails. In 1974, France celebrated the twenty-fifth anniversary of *le Comité National de Sentiers de Grande Randonnée* (The National Committee for Long Distance Footpaths) and over 6,000 miles of marked paths. Holland boasts one of the world's most extensive bicycle, equestrian and hiking trail systems. Under the National and Scenic Trails Act, the United States has developed the Appalachian and Pacific Crest long-distance hiking trails.

A few agencies have concentrated on individual trails in Alberta. However, to date, there has been little in the way of an organized study to preserve the province's historic trails. The apparent consensus expressed by outdoors enthusiasts is that there just aren't enough safe and enjoyable trails in Alberta. *Historic Trails Alberta* is dedicated to the rediscovery and, unblushingly, the promotion of Alberta's historic trails. It is not, however, an invitation to abuse our habitat or the fragile remnants of a living past. Respect for private property and common sense are required of anyone who enjoys the great outdoors.

The First Trailblazer — Anthony Henday

It was the fall of 1754 when Hudson's Bay Company envoy Anthony Henday came to Alberta. Save for the symphony of a fiery autumn, he met no fanfare — no dipping banners, no salvos of artillery, no pomp or regalia. And yet, as history has demonstrated, Henday's was a fateful coming. Not only was Anthony Henday the first recorded white man to enter Alberta, but he was the first to behold the glorious Rockies north of Mexico, the first to meet Alberta's Blackfoot Indians and the first to trade directly.

In prior decades, the Hudson's Bay Company had restricted its commerce to York Factory on the Hudson Bay shore. During long yearly pilgrimages, the Crees and other tribes took their furs to the fort. The Hudson's Bay traders reaped the harvest of a prairie market they had never seen. Meanwhile the French coureurs-de-bois from Montreal, under economic pressures, expanded westward in quest of trade. Their scheme was to pre-empt the Indian freight bound for Hudson Bay along the northern rivers, the conduits of transport. The Hudson's Bay Company, piqued by the efforts of the interloping French, saw that a bit of market reconnaissance was called for.

It was Anthony Henday, a smuggler turned trader, whom York Factory Chief Factor James Isham chose when he sized up his men. As it turned out, Isham couldn't have made a better choice.

Who better than Henday — a robust man, skilled woodsman and experienced navigator from his days at sea — could impress the Indians and command their respect.

Henday kept a journal which though stayed with the prose of a sober hand, accurately logged the course of his journey.

> Level land, few woods and plenty of good water . . . Indians killed 8 Waskesew [Elk]

Was this Henday's first impression of shining Alberta? No doubt his interest lay with the promise of business, and yet being a man of spirit with a thirst for adventure, he was surely inspired by his marvelous exploits. Henday came as a goodwill ambassador to befriend the mysterious Blackfoot Indians and woo their trade. In the process, he achieved the first reconnaisance of the heart of the continent.

Near present-day Irma, Henday encountered an Assiniboine

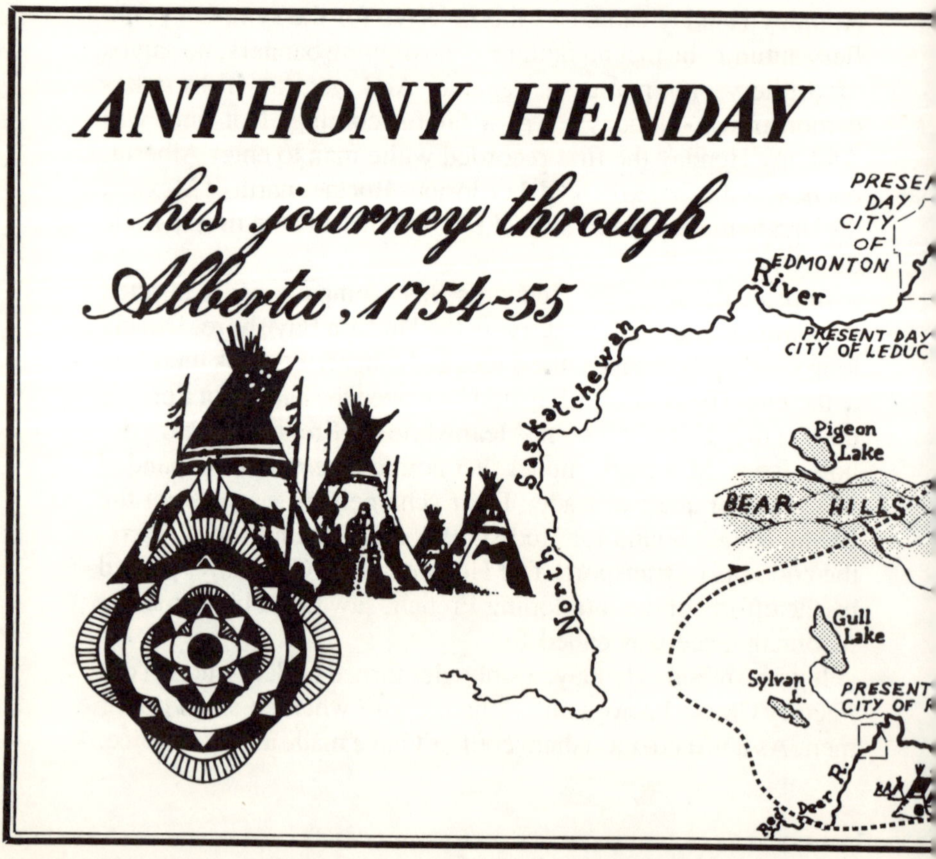

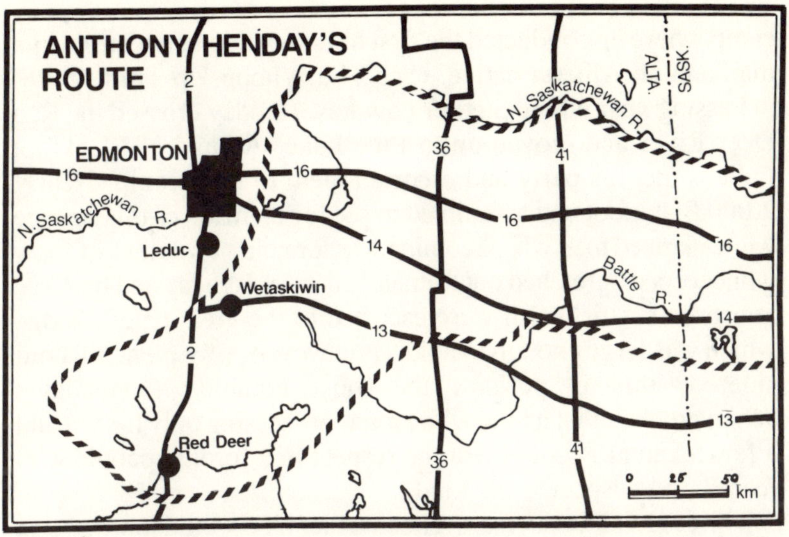
ANTHONY HENDAY'S ROUTE
EDMONTON
Leduc
Wetaskiwin
Red Deer
N. Saskatchewan R.
Battle R.
ALTA.
SASK.
2
16
14
13
36
41
0 25 50 km

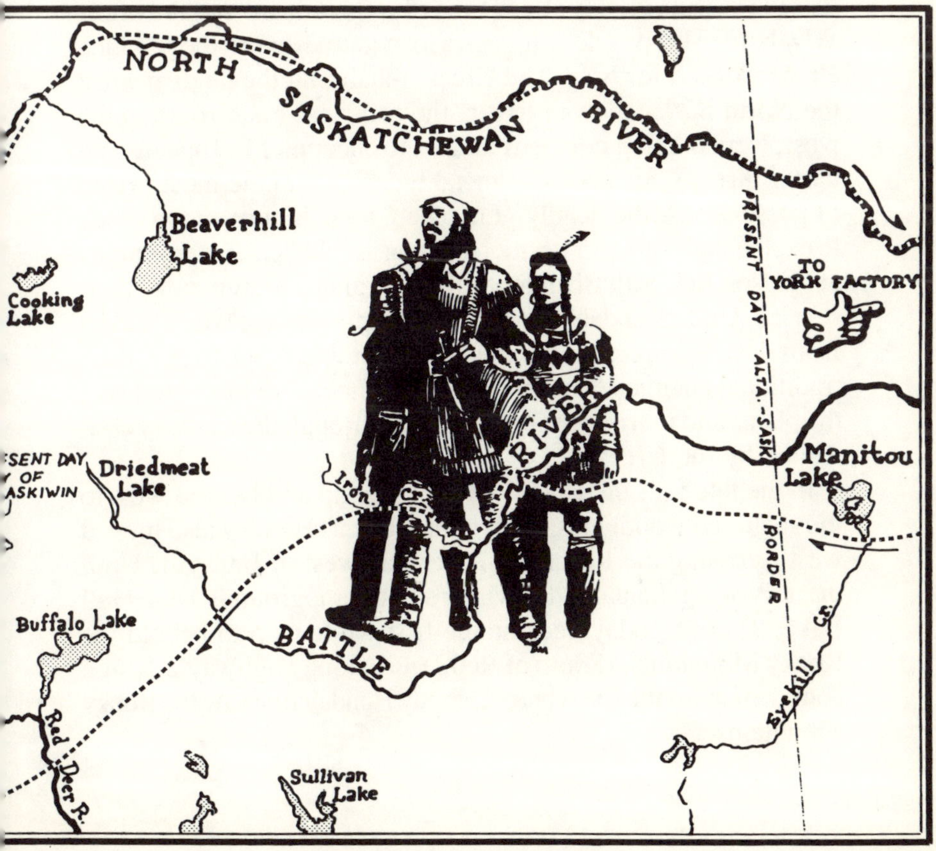
NORTH SASKATCHEWAN RIVER
Beaverhill Lake
Cooking Lake
TO YORK FACTORY
PRESENT DAY ALTA.-SASK. BORDER
Manitou Lake
Driedmeat Lake
Iron Cr.
BATTLE RIVER
Buffalo Lake
Red Deer R.
Sullivan Lake
Eyehill Cr.

camp where he conducted the first horse trading between a white man and an Alberta native. (See The Whoop-Up Trail.)

Passing near where Stettler now lies, Henday crossed the Red Deer River and moved on to Pine Lake. A band of Blackfoot Indians met his party and escorted them to their camp. Nearly 2,000 Blackfoot had assembled to see for themselves this strange white-skinned man whose coming was foretold. Two rows of snow white teepees stretched out for nearly a mile. Henday and his Cree interpreter Attickasish were escorted to the great chief's lodge which was large enough to hold a powwow of fifty men. All considered, this is perhaps the most notable of Henday's achievements: that a lone white man, possessing only his wit and a few token gifts,could win the respect of a foreign people who feared and suspected his kind.

Thrice Henday asked the reticent chief if he would journey east to trade at Hudson Bay. Each time, the chief dismissed the offer. Why burden his people with such a long journey when the peddling Crees worked for him. The Crees, inhabiting the lands nearest the North Saskatchewan River, the principal trade route to the west, were the first northern Indians to encounter Europeans and barter directly for goods — metal utensils, and the most prized of possessions, the deadly gun. The Crees, in turn, were quick to exploit their superior position. Acting as middlemen, they traded at Hudson Bay with Blackfoot furs, charging a premium for their service. Understandably, they strove to prevent the white traders from dealing directly with the Blackfoot and other tribes, their traditional enemies. The Blackfoot, in turn, were loathe to leave the plains and journey the long distance through the territory controlled by the Crees.

In the late fall, the Blackfoot broke camp and headed west to the sheltered woodlands, their winter haunt. Henday also moved west, crossing the Red Deer River just west of Innisfail. Until the following January, he wintered at campgrounds near Gull Lake. There Henday became the first white man to behold the Rocky Mountains. (South of Red Deer, along Highway 2, a sign commemorates the site where Anthony Henday first saw the Rocky Mountains.)

Henday's Route

Setting out from York Factory at the mouth of the Hayes River on Hudson Bay, Henday and his Cree guides paddled the few kilometers around Marsh Point and ascended the Nelson and Saskatchewan rivers to The Pas (in modern-day Manitoba), from where they portaged to the Carrot River and sailed upstream until they reached the country just northeast of today's Humboldt, Saskatchwan. From Humboldt, they continued to Saskatoon where they crossed the South Saskatchewan River and struck out northwest towards the North Saskatchewan, following its banks through the Eagle Hills north of Wilkie and towards Cutknife. Travelling south of Manitou Lake, they then veered northwest across the Battle River near Wainwright. From Wainwright their journey led them north of the Battle River to Killam. They crossed the river again between Forestburg and Donalda. Henday and his guides continued southwest past Stettler, crossing the Red Deer River east of Delburne. Somewhere between there and Pine Lake they met the encampment of Blackfoot Indians.

When the Indian camp broke for the winter, Henday's entourage departed westward to re-cross the Red Deer River west of Innisfail. They spent the winter of 1754-55 camping and hunting in the forests as far west of the modern Calgary-Edmonton Highway as the Clearwater River. In early spring, they worked their way through the Bear Hills and travelled along a path east of the line between Edmonton and Wetaskiwin. Finally, they struck the North Saskatchewan River at a point just below Edmonton and followed it down to the mouth of the Sturgeon River.

There they camped while they made canoes and waited for the ice to break up. When the ice loosened, they sailed the Saskatchewan River to The Pas and retraced their steps to York Factory. The journey had carried Henday a total of 2,500 miles in 52 weeks.

Points of Interest Along Henday's Route

In the 1870s, Tail Creek was a Metis village said to have been the largest west of Winnipeg. Near the Tail Creek area on Buffalo

The Cooking Lake Trail in 1902. As he headed for the North Saskatchewan River, Henday passed through the Cooking Lake parkland.

Lake, a legendary Indian cemetery is believed by some to entomb the dead of centuries past. The desecrating swells of the river still unearth skulls and bones.

In 1874, during an annual Metis hunt, an untold many were besieged and killed on the plains by a disasterous storm. Later the smallpox epidemic swept the plains and killed half of the Tail Creek population. The plague came during the hunt, turning their teepees into palls of darkness. Wrapped in hides, many were buried at Tail Creek in graves covered by traditional, ornate Metis tombs — wooden structures with symbolic etchings and roofs.

Approximately eight miles northwest of Stettler at Boss Hill, where the Indians obtained flint for their tools, one can still find the occasional artifact. Though Buffalo Lake is shaped somewhat like a buffalo with its head lowered, it more likely is so named because of the many buffalo that once roamed its shores.

Nearby, the Haunted Lake just east of Alix, is also steeped in mystery. Legend has it that when the moon is bright during winter nights, the head of an elk can be seen protruding from the ice in the middle of the lake. It seems that some hunters from a nearby Cree camp discovered the carcass and rushed onto the ice to cut it out. But before they could reach it, the head began to quake, the ice broke and all perished in the cold depths. Henceforth, the Crees avoided the lake, said to be haunted by their dead brothers.

The night before entering Alberta, Henday camped just north of Chauvin near Sherlock Lake. In the early days of settlement, a trail to Battleford passed close by. It began at Sounding Lake and the Neutral Hills, skirting the Battle River near the mouth of Blackfoot Coulee. From there it proceeded to old Fort Pitt. This trail, like many other Indian trails, is believed to have been hundreds of years in the making.

Due south of Ribstone Creek, nestled in the Neutral Hills, Sounding Lake, too, is steeped in legend. Long ago — so long that only "Old Man" knows the truth — Mikisew, the Great Eagle, emerged from the water, grasping a monsterous snake in his talons. The bird-god's wings roared in flight like thunder across the hills and open prairies.

The Neutral Hills offer a magnificent view of the surrounding ranchland, laced with sapphire lakes and the green of wooded bluffs. A plaque on the highest point of the Neutral Hills proclaims the ground — where once a legendary battle raged — "neutral" by divine providence. To ward off further bloodshed, The Great Spirit (in the lull of battle) molded the hills as a barrier between the two rival Indian tribes. An abundance of arrowheads and artifacts of war have been discovered nearby, and the stones of graves are occasionally disturbed.

Scholars argue, however, that the hills were deemed "neutral" because the Blackfoot and Cree, who hunted the teeming buffalo in the area, confronted each other in ongoing battles, but finally came to the practical agreement that it would be best for all concerned if they made peace and shared the ample wealth provided.

By September 24, Henday had reached Iron Creek northwest of Hardisty. Iron Creek is so named because of the chunks of bog iron found nearby. It was also once home to a large meteorite, the "Manito Stone." We do not know the exact date when the meteorite roared through the heavens. Perhaps it was very long ago and simply exposed by erosion on the hill overlooking Iron Creek. But for generations, Iron Creek tribes revered the Manito Stone at which they left sacrifices in return for good hunting, health and prosperity. In 1869, John McDougall discovered the meteorite and had it moved to the Victoria mission. Though white men have scoffed at such superstition — the celestial rock being to them a mere curiosity, a manifestation of a natural order — the Indians' fears were seemingly confirmed the next year in 1870 when the smallpox epidemic decimated their people, the buffalo herds thinned and starvation swept the parklands. Canada's largest meteorite is now displayed at the museum of Victoria College in Toronto, Ontario.

Further west, Henday crossed the Battle and Red Deer rivers, where the hand of the ages displays its craft. Here, layer after layer of exposed strata — petrified sea creatures, agatized trees and coal seams — reveal the story of creation. Though the landscape here is not as spectacular as the badlands further south, Henday must have been impressed by the mystical aura of the alien landscape. How inspired by the all of Alberta's untouched beauty was this giant of his time when he wrote: "I cannot define the fineness of the weather, and the pleasant country I am now in."

Wonderous were the many achievements of such trailblazers as Anthony Henday. Today a comfortable, mechanized cross-country jaunt pales by comparison. We are an impatient society — we move in the swift current of technology. Even with our flawless maps, to somewhere lose our way, to interrupt for the slightest moment our hectic pace, causes us great anxiety, even fear.

Surely Henday and others of his determined mold were fearful in the knowledge that they were treading a domain unknown and potentially dangerous, but what matters is that their stamp on the west helped forge a great nation. For this, they will always be remembered.

Trails From The East . . .

The West's earliest trails were etched across an elemental landscape teeming with bufflo. A retreating ice age had left pasture land, birthing perhaps the largest herds the world has ever known. Seasonal hunts led the Indians into the southwest prairies of the Missouri, the Souris and the Grand Coteau rivers. And the hunt was of economic, social and religious significance. The radiant sun god blessed the land. The buffalo were manna — lesser gods, icons of faith.

When Europeans stepped out into the inland sea of grass, the great animals were perceived as a provisioning source. They became a commodity in the new corporate order. The buffalo sustained a people who did not revere its celestial qualities. For over a century, fierce competition pushed the fur trade west. Traders, having little time to trap or hunt, were perilously distant from the eastern posts, their source of provision. Buffalo meat provided by the Indians sustained their lives.

By the early 1800s, trading companies controlled an empire from Lake Superior to the Pacific Ocean. It was a business which called for as many able-bodied men as could be inducted to duty. Hunting buffalo for trade became the chief occupation of the eastern Plains Indians of the Souris and Qu'Appelle regions. Soon, white traders — freeman voyagers whose Hudson's Bay Company contracts had expired — settled the area and joined these tribes, learning the secrets of the disciplined hunt, which they taught to their progeny of mixed blood, the Metis. The Metis adapted to the traditional hunt with guns and improved transport and so mastered it, some would argue, that the buffalo herds were slaughtered to near extinction.

For the first time, the hunt was organized and the Metis people, more closely associated with the white traders than their pure blood relatives, worked for wages.

The hunting parties gathered at the missions near Fort Garry or St. Francois-Xavier and made their way along the Pembina Trail running just south of the Assiniboine River to the Riviere Iles de Bois and the Pembina Hills. The Pembina cart track was a branch segment of the freighting route from Fort Garry to Edmonton via Fort Carlton and Victoria (later Pakan) in Alberta.

Through the prehistoric delta of the Assiniboine River, the Pembina Trail wound its way through the "Bad Woods," over the slopes of the Sand Hills, through clusters of spruce and dunes of the Dry Dance Hills and along the Souris to its junction with the Assiniboine River. This segment of the Winnipeg-Edmonton trail would later be named after Yellow Quill, an Indian Chief who was rebellious to the treaties which forced his people on to reserves. The Yellow Quill Trail was a portage link with the Upper Souris and Missouri River basins and with Lake Winnipeg. Early French explorers and Hudson's Bay Company fur traders used this trail and established posts along the Assiniboine River and at the mouth of the Souris.

While the trade era boomed, work was plentiful, but with time the fur supply was exhausted. Because the Hudson's Bay Company could no longer support the Metis, the hunt became their means of subsistence. By 1850, cart brigades lurched and shrieked in growing numbers over the prairies to and from settlements now increasingly supplied from the United States. Much to the anxiety of patriots loyal to the British Crown, this U.S.-Red River commmercial link signified unwanted change.

The old order, a compromise of river-lot farming, buffalo hunting, and old-world connections, was being brushed aside. The forerunners of an gricultural age, farmers, land-hungry speculators and merchants invaded the vast and unclaimed Dominion. And the Hudson's Bay Company, daunted by its leviathan empire, accommodated a new Canadian political power and opened the way for settlement west.

For two centuries, the great northwest had been thoroughly explored for the fur trade. What was still lacking was the scientific detail on the climatic and geographical features needed if organized farming was to succeed. For the first time, scientific expeditions, such as those led by Hind and Palliser, struck out on fact-finding missions. We needn't recount Palliser's or Hind's exact course — as both used trails discussed in this book. From the St. Lawrence seaway to the Pacific coast, they explored the waterways and overland trails by way of cart, horse, snowshoes and dog-team transport. They explored the basin of the Great Lakes and the unpeopled prairie sea, ancient ridgeways and Indian trails all drifting in random fashion over the parkland and lowlands toward the Red River settlement.

Eventually the rich black soils of the Manitoba and Saskatchewan prairies were well advertised. A whirlwind of "Manitoba fever" brought immigrants from around the world to populate the wheatlands. Between 1876 and 1881, some forty thousand people—Icelanders, Mennonites, French and English tenant farmers—swarmed to Canada. Shacks adjoined sod cabins and immense plowed fields spread over the prairie.

While few Manitobans trickled south to the midwestern United States, colonization parties from Michigan and Wisconsin came to Canada to settle the land unclaimed by the Red River colony. The towns of Ste. Agathe, Morris, and Emerson were founded along the St. Paul-Winnipeg Trail (today Highway 75 running south from Winnpeg). In turn, Emerson and Morris were major points of dispersion for settlers pushing west.

Western settlement followed a similar pattern of congregation around wooded bluffs and abundant water. All hoped to be located near the transcontinental railway. Few pioneers, however, would dare to challenge the isolated prairies. Settlers striking out from Emerson or Morris followed two lines of approach to the wooded ranges of the Pembina Escarpment and to the small settlements collectively known as Pembina Mountain, united by the Pembina Trail.

Other Manitoba settlers, English half-breed hunters turned farmers and grazers under the wing of the church, populated settlements threaded by the Saskatchewan Trail and branch roads. Many of these villages died away, but some have survived those early days of rebellion and divided loyalties.

Grand Marais on the southeast shore of Lake Winnipeg grew into a permanent settlement. Southeast from St. Boniface, where the Seine River emerged from the ridges and forests of the eastern highlands to the land of meadows and oak groves, the Metis founded Oak Point and the mission of Ste. Anne des Chènes. North of the Whitemud River, just removed from the Carlton Trail along the North Saskatchewan River, the fishing village of Westbourne came into being.

To expedite settlement of the Northwest, the government appointed S.J. Dawson to cut a road from St. Boniface to the North West Angle of the Lake of the Woods. From the lean soil of eastern Manitoba's Precambrian Shield, hundreds of families upheaved their plowshares and migrated west over the Dawson Road, founding such settlements as Prairie Grove and Clear Springs near Winnipeg. For the poor and destitute, it was a struggle to survive those early years plagued by grasshoppers, swept by fires, withered by drought. The Canadian Government set up relief services in local parishes to distribute food and to succor the needy.

This eastern Manitoba region, the Portage area, was squatters' territory. It was unspoken law that a man could lay claim to as much land as he could physically manage. However, the white man's law did not apply to the resident Metis and when the zealous pioneers overran their territory, trouble was inevitable.

Everywhere, unscrupulous white men were buying up Metis titles with a few bottles of whiskey. The land barons acted with the belief that the newly formed Canadian Government was too weak to intercede. The government only confirmed such notions by convincing itself that the Metis spokesman Louis Riel would not incite his people to battle. As generations of Canadians have since known, the government was wrong.

But to this point Alberta was estranged from the expansion fever. "Though more than a century has passed since the first white man has tramped through Alberta, neither he nor his successors had laid hands on the province's natural beauty," states James MacGregor in his *A History of Alberta.*

More than 100 years after Anthony Henday became the first European to enter what is now Alberta, the region was sparsely settled. Hudson's Bay Company outposts at Fort Chipewyan, Dunvegan, Edmonton and Rocky Mountain House controlled much of the northern economy, while the American whiskey traders held fortified posts at Whoop-Up and Standoff. Early missions at Victoria (Methodist), St. Albert and Lac Ste. Anne were largely Metis and native. The powerful Blackfoot Confederacy still controlled the plains, though the decline of the buffalo and the inevitable surveys for the railroad numbered their days. In 1870, perhaps two dozen permanent white settlements existed in Alberta. By 1885, thousands of immigrants were pouring in on the newly opened railroad.

The Red River Buffalo Hunt

The traditional Metis buffalo hunt was a grand spectacle. One can picture the skilled huntsmen as they wheel their mounts upwind of the herd. The captain signals and the brigade charges, spreading out across the plain, hooves pounding, nostrils flaring, sinewy muscles taut. The alerted bison flee and the warriors bear down with deadly accuracy, their guns leveled over the necks of their steeds. Having no time to pause, each marksman carries a pocket of loose powder and a mouthful of iron shot. As he rides, he quickly reloads — ramming power and spitting shot down the barrel of his gun — and pounds the stock against his thigh to pack the mixture.

The dust clears and the Metis hunters turn their foam-flecked runners back towards camp. They leave behind a scattered mass of still-kicking behemoths. Each hunter marks his kills by dropping a personal article — a glove or hat.

When the killing is finished, they return to camp where the women prepare the meat, stripping the carcasses of all useful parts — the hides for robes, the humps for fat, and the tongues to be pickled as delicacies for special celebrations. The remains are left for coyotes, crows and scavenging varmints.

The dressed meat is loaded onto the train of Red River Carts—hundreds of them. The riches will be sold, bartered or eaten, to support the Red River Metis colony for another year.

The Red River Cart: Charrette de la Riviere-Rouge

Invented at the turn of the nineteenth century, the Red River cart was neither handsome nor stylish. It was an austere conveyance, rugged and true at a time when prairie transport was sorely limited.

The strongest pack pony could carry only about three hundred pounds. The canoe, though swift in water, could not sail the prairie sea. The iron carts of Quebec were ideal, but metals were scarce in the west. The Red River cart solved these problems.

The Red River cart was constructed of wood — not one scrap of iron, not even a nail. It was crudely formed: a wooden box fastened to a makeshift axle, usually a squared-off tree trunk with tapered ends. Its wheels were made either of disks of solid wood or of several pieces which were curved and dovetailed together to form the rim and spoked to a central hub. Two short poles, lashed to the box, served as shafts. The whole contraption was held together by wooden dowels and "shaganappi," buffalo rawhide which shrank and hardened as it dried. Rawhide also lined the wheel rims to soften the ride. When repairs were in order, a driver could procure materials from a nearby tree or herd of buffalo.

But the Red River cart did have one flaw — and an annoying one. With its wooden hubs constantly scraping the bone-dry axles, the cart would wail and screech as though protesting its heavy burden. (No need to put your ear to the ground to fortell the coming of this wagon.) Even the Metis scoffed at the very cart which sustained their lives. It was said by the Metis that the buffalo had left the plains in flight from the terrible noise.

The Metis used the Red River cart to transport buffalo meat and pemmican (dried fruit and meat) to the fur traders wth whom they bartered. Until the advent of the Red River cart, the buffalo hunters were limited in the amount of meat they could kill and transport. However, by the middle of the nineteenth century, a single hunting expedition might have involved as many as 1,500 carts and produced over a million pounds of buffalo meat.

Through violent storms, fires and quagmire the Red River cart endured the rigors of prairie life unitl, with the arrival of the steamboat and iron horse, it faded into the dust of its beaten trail.

The Carlton Trail

When people speak of the old Carlton Trail, they are referring to what is possibly the oldest westward overland trail in Canada. It was Dominion Canada's "Oregon Trail," the beaten path linking east and west, crossing the Red River in the heart of the continent, passing through virgin prairie and parkland, Indian haunts and the future provinces of Manitoba, Saskatchewan and Alberta.

The Carlton Trail was named after an important Hudson's Bay Company post located on the North Saskatchewan River, midway between North Battleford and Prince Albert. This trail is known through historic documentation as the Victoria Trail, Winnipeg Trail, or Fort Pitt and Saskatchwan Trail, depending on the location of its different segments. For example, the segment from Edmonton to Victoria (now Pakan) is referred to as the Victoria Trail, and that which lies from Victoria to the eastern Alberta border is the Winnipeg Trail

The original route of the Carlton Trail came into being with the amalgamation of the Hudson's Bay Company and the North West Company in 1821 and the subsequent founding of Fort Pelly as headquarters in the Swan River district in what is today Saskatchewan. Even then, much of the trail's general path, though dimly marked in places, was familiar terrain to fur traders who roamed west from their posts on the upper Assiniboine River as far and wide as Touchwood Hills and the Quill Lakes in central

Saskatchewan. For generations the annual buffalo hunts had carried the Red River Metis west into the plains along a network of well-beaten trails leading from Fort Garry to Portage La Prairie. One such trail was a branch heading towards the Touchwood Hills, the rough-hewn vestige of which was destined to become Canada's most important transcontinental road.

Between Fort Pelly and Carlton House, however, there were no connecting waterways to guide travellers through the isolated countryside to the Saskatchewan River. Moreover, it was treacherous going when treading the no-man's land between the territories of the friendly Cree and their blood-sworn enemies, the fierce Blackfoot and allies of the Blackfoot Nation.

Despite such adversity, Hudson's Bay Company Governor George Simpson, during his journey to the Columbia District in 1824, braved this dim passage when few men would dare. Accompanied by Chief Trader James McMillan and eight company employees, Simpson and party, despite rumours of ambush, rampaging floods, and the grave threat of becoming lost, succeeded with gusto. Later, during his famed trip around the world in 1841, Simpson used the trail as we know it. Simpson's courageous example demonstrated the feasibility of extensive overland travel at a time when the growing importance of the Red River District as an administrative center increased the need for swift travel and communication. Within a decade of Simpson's trailblazing journey, the Carlton Trail developed into an important link of a tried-and-true transcontinental trails network.

By 1860, however, an extensive railroad network in the United States proved a cheaper and more efficient means of transport than did transcontinental overland freighting by wagon. Red River was evolving into a vital community and consequently usurped a great deal of the commerical freighting to the Swan River and Saskatchewan Districts.

Yet, for the next two decades the Carlton Trail was the only well-tread Canadian overland route between Fort Garry and Edmonton. Beginning from what is now Portage Avenue in modern Winnipeg, this historic trail stretched for 900 miles through Portage La Prairie, Minnedosa, Shoal Lake and Fort

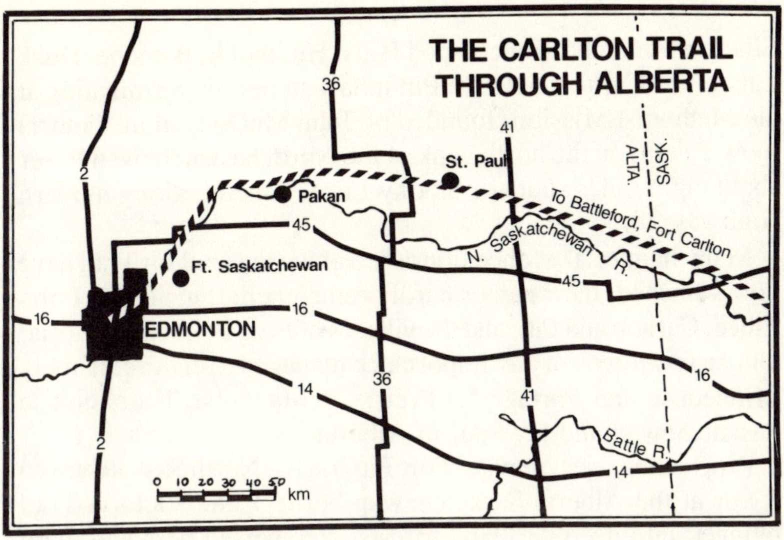
THE CARLTON TRAIL
THROUGH ALBERTA
St. Paul
Pakan
To Battleford, Fort Carlton
N. Saskatchewan R.
Ft. Saskatchewan
EDMONTON
ALTA.
SASK.
Battle R.
2
36
41
45
16
14
0 10 20 30 40 50 km

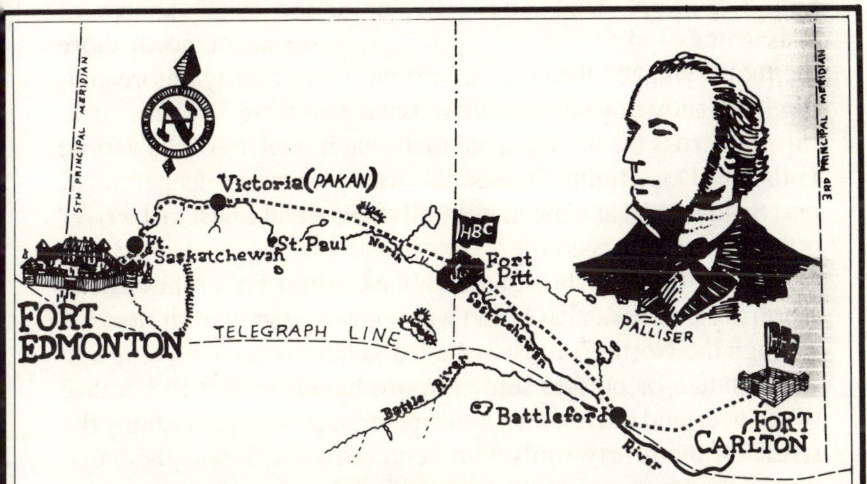
5TH PRINCIPAL MERIDIAN
3RD PRINCIPAL MERIDIAN
Victoria (PAKAN)
Ft. Saskatchewan
St. Paul
North
HBC
Fort Pitt
Saskatchewan
FORT EDMONTON
TELEGRAPH LINE
PALLISER
Battle River
Battleford
River
FORT CARLTON
CARLTON TRAIL
CIRCA 1880
D. MARKO 85

Ellice in Manitoba, Touchwood Hills, Humboldt, Batoche, Duck Lake, Fort Carlton and Fort Pitt inSaskatchewan, terminating at the Methodist Mission, founded by John McDougall in Victoria (now Pakan) on the north bank of the North Saskatchewan River about eight miles south of Smoky Lake, Alberta, along modern Highway 28.

Many hamlets that once flourished along the Carlton Trail have now vanished, their names hardly remembered, including Forts Ellice, Carlton and Pitt, and the villages of Frog Lake and Victoria. Only a few towns of any importance managed to endure, namely Minnedosa and Portage La Prairie in Manitoba, Humboldt in Saskatchewan and St. Paul in Alberta.

Proceeding west, leaving Fort Pitt (on the North Saskatchewan River at the Alberta-Saskatchewan border), the Victoria Trail plunged into deep parkland forests interspersed here and there with fire-swept dead zones. It skirted the sandy shores of Wasagamus Lake (Cree for clear water), on whose north shore the future village of Frog Lake would thrive briefly before ending in tragedy during the "Frog Lake Massacre." Then, striking out across the wide glacial melt-water coulees of the Moose Hills and Dog Rump Creeks, the trail penetrated deeper west, past Saddle Lake and the Snake Hills to descend at last to the river at the Victoria mission.

Much of the Alberta segment of the Carlton Trail followed the North Saskatchewan River and was known as the "North Route." Though the North Route is believed to have originated long ago as an Indian or buffalo trail, it more likely developed because of the overland movement of fur traders between posts along the river. Of the hearty souls who ventured west, it was the "free traders" who perhaps best exemplified the pioneer spirit, rising to the back-breaking task of nation building. With the outcome of the Sayer Trial at Red River in 1849, there was set a precedent in that from that date forth, free traders could enter into competiton without reprisal from the Hudson's Bay Company in the barter for furs. With traders vying fiercely for potential prairie wealth, the development of trails and settlements spreading westward was inevitable.

John A. McDougall was one of the first and most renowned of the Alberta free traders to trek the Carlton Trail in his commercial ventures. Though McDougall was admirably successful in the closing of deals and the making of money, he appreciated the finer qualities of life on the trail, of which he wrote in his memoirs:

> It was a healthy life, a free one and a most fascinating although a lonely one at times, travelling across the open country in those early days. But we enjoyed the care-free outdoor life and learned to admire the beauties of nature, appreciate the vastness of the universe and had our belief confirmed in the Great Director above who watches and rules over all.

Later when Rev. George McDougall and his son John established the Victoria Mission in 1863, the Carlton Trail entered into a new phase. Where once the trail only accommodated pack horses, it now was capable of handling carts and wagons. The Winnipeg Trail became a vital freight route. Father Lacombe deserves much of the credit for the trail's development as he claimed to have employed the first cart brigade to carry freight over the trail from Red River to Carlton House in 1862.

Victoria, the first permanent settlement between Fort Pitt and Edmonton along the trail, administered the faith to the nearby Cree Indians. In 1864, George McDougall opened the first protestant schools west of Manitoba — one at Victoria, and the other north at Whitefish. Two other missions were also established near the Victoria settlement, including St. Paul de Cris, a Roman Catholic mission founded by Father Lacombe southeast of Victoria along the North Saskatchewan River near present-day Brosseau. George McDougall went on to found a church in Edmonton and at Morley, west of Calgary. Victoria was later renamed Pakan in honour of Cree Chief Pakan, a friend of the McDougall family.

Three of the family — Mrs. McDougall and two daughters — died during the smallpox epidemic of 1870 and are buried near the river on the west side of the present-day Victoria compound, an historic site developed for visitors by Alberta Culture.

During the height of the fur trade in Alberta, the north route along the North Saskatchewan was the trail preferred by traders penetrating the west. Between Fort Edmonton and the eastern Alberta border, no less than twelve trading posts lined the north bank at one time, affording the only protection against attack by the Blackfoot tribe gathered to the south. Consequently, though the south route or south trail was considerably shorter and less demanding than the north trail, its use was inhibited for many years before the west was tamed.

The Carlton Trail was significant in the initial planning of the Canadian Pacific Railway when it was proposed as the route across the Northwest to the Yellowhead Pass in the Canadian Rockies. Stanford Fleming, engineer-in-chief of the Canadian Pacific Railway, selected survey party leaders to investigate possible routes while he personally surveyed the Carlton Trail in 1872. By then the trail was well established and from Fort Pitt to Edmonton he travelled by buckboard.

It was then believed that the Canadian Pacific Railway would begin at Selkirk, Manitoba, north of Winnipeg, and proceed northwest over the narrows of Lake Manitoba, following the Carlton Trail west to Edmonton and through the Yellowhead Pass. While the railway was being constructed, free traders hauled materials over newly cut roads and old portions of the Carlton Trail. In the early stages, before the planned route was altered, Victoria mission was an important trading post from which the survey crews secured supplies. However, after the railway was built in the south, Calgary soon eclipsed its commercial potential and the Hudson's Bay post at Victoria was closed.

The first mail run over the trail was conducted along the lake route between Fort Garry and Fort Livingston in 1874 by John Matheson who travelled by dog sled. The service was extended to Edmonton two years later. By1876, regular cross-country mail service was conducted along both segments of the Carlton Trail. When a shorter route south of the Saskatchewan River was adopted, the Carlton Trail was used casually by local traffic. The advent of sternwheelers plying the Saskatchewan River, and later the railroad to Calgary, likewise eclipsed the trail's importance as a freighting route. From the inaugural voyage of the first steamer

PROVINCIAL ARCHIVES OF ALBERTA–
E. BROWN COLLECTION

A major centre in Alberta before the arrival of the railroad was Victoria (now Pakan), just south of Smoky Lake. A Methodist mission and trading post, Victoria is pictured above in 1885.

en route to Edmonton from Lake Winnipeg, it was evident that the era of York boats, dog trains and Red River carts was destined to end. From Fort Garry to Edmonton the ship, carrying some 130 tons — the equivalent of 325 Red River carts — took just 34 days to complete its trip. When conditions permitted, the trip by cart from Fort Garry to Fort Carlton alone took about 25 days.

To travel the Carlton Trail, head northwest from Edmonton on Highway 15 to Highway 825, then north. This route passes a sign commemorating Fort Augustus and Fort Edmonton. Albertans generally believe that Fort Edmonton was originally built just below the present Legislative Building in Edmonton, but, in fact, the Hudson's Bay Company and the North West Company first built their forts in 1795 near the mouth of the Sturgeon River (near present-day Fort Saskatchewan) and moved the forts a number of times until, after the companies merged in 1821, Fort Edmonton came to rest at its final location in 1830.

The road between Edmonton and Victoria (Pakan) only approximates the Carlton Trail. The actual trail hugged much closer to the north side of the Saskatchewan River.

In Waskatenau, take the bridge leading south. Running in an easterly direction, a gravel road follows the original trail and crosses Highway 855 to lead you to the Victoria Settlement Historic Site, where a replica of the old mission complete with many artifacts is open to tourists.

North of the Victoria settlement, a meandering sandy road approximates the old Carlton Trail in spots revealing splendid views of the North Saskatchewan River valley. This road crosses the White Earth River and then turns to pavement as it passes through the Saddle Lake Indian Reserve.

The Carlton Trail then passes through Brosseau, nearby which Father Lacombe founded the St. Paul des Cris Mission. Father Lacombe, having anticipated the fate of the natives, had hoped to turn them from nomadic hunters to people of the soil. Although the community failed, there remains a quaint little church which overlooks the river.

At St. Brides at the junction of Highway 36 and 28, a cairn marks the original Winnipeg segment of the Carlton Trail.

The original trail continued through Elk Point, just south of which stood the rival fur posts Fort George and Buckingham House. The contest for furs was so intense that by 1795, the beaver of the area were all but depleted. In 1800, the forts were abandoned for the new Forts Edmonton and Augustus. In Elk Point, Steve Andrishak owns a private museum in which he displays artifacts from Buckingham House (whose site is now an historic point of interest).

The Carlton Trail leaves Elk Point and continues east, passing close by Riverview and onto Frog Lake where a cairn and seven graves mark the site of the tragic Frog Lake Massacre. In Dewberry on Highway 45, the Dewberry Valley Museum displays items from the massacre.

Along the highway running south from Frog Lake, the Carlton Trail leads you to the site of old Fort Vermilion (Paint River House) founded by the North West Company. This is now an historical

DAVID CUNNINGHAM

The cemetery at Frog Lake. During the Northwest Rebellion in 1885, Big Bear and his band of Crees attacked and killed nine people at the Frog Lake Indian Station. The Carlton Trail passed by the lake's south end.

site and provincial park.

To continue along the original Carlton Trail, travel east along Highway 641 until it veers south, then take the east road into Saskatchewn through Onion Lake and southeast towards the site of old Fort Pitt.

Fort Pitt was established by the Hudson's Bay Company in 1835 and had an auspicious lifespan of about fifty years until it was destroyed by Indians. The site of the old fort is now a picnic and campground with a stone monument detailing the area's history.

Leaving Fort Pitt, a few roads lead to Frenchman's Butte, just north of where Canadian troops and Big Bear's Indian band confronted each other following the destruction of Fort Pitt. Along the way from Frenchman's Butte and Battleford, north of Edam, a marker set in the shade on a picnic ground identifies the location of the Carlton Trail.

South of Meota, Highway 40 duplicates the trail's route. It veers east away from the North Saskatchewan River and brushes the southern edge of the Thickwood Hills. The original Carlton Trail continued eastward by crossing the river east of Blaine. To cross the North Saskatchewan today, proceed north to Wingard and take the ferry. Fort Carlton is just a few kilometres south of where the ferry docks.

Fort Carlton, situated in the river valley, was destroyed by fire in 1885 and eventually rebuilt. It is now the site of archeological study and historic exhibits. Close to the fort is a picnic area and campground. Continuing east from Fort Carlton, the Carlton Trail wound through the river valley and proceeded to Duck Lake. This road passes a monument that identifies the site of a battle between the North West Mounted Police and the Metis in 1885.

Continuing from Duck Lake, a road follows the trail by St. Laurent. During the 19th century, St. Laurent was a provisional base for the Metis settlement in the north. The trail has now arrived at the South Saskatchewan River. High on the west bank near St. Laurent, a log shrine is dedicated to Our Lady of Lourdes.

The Carlton Trail then crossed to the other side of the South Saskatchewan to the settlement of Batoche. Batoche was the nickname of a Metis merchant otherwise known as Xavier Letendre. Mr. Letendre operated a ferry service in 1871 that competed with a Hudson's Bay scow further upstream. A sign at the crossing details the history of Batoche.

Leaving Batoche, the Carlton Trail ran southeast, passing such small hamlets as Wakaw, Cudworth and Humboldt south of Dafoe on the Salt Plains. Two Lakes near Dafoe, Big Quill and Little Quill, were once the nesting grounds for thousands of geese. Traders sold goose quills for writing nibs.

Highway 15 approximates the trail as it went southeast to Touchwood Hills. Touchwood Hills lie between Punnichy and Lestock and was founded as a trading post of the Hudson's Bay Company in 1849. This post was a hub of regional activity, through which numerous other old trails passed, including a southwest trail that ran to Last Mountain House, a southeast trail to Fort Qu'Appelle and Wood Mountain, and a trail going east to Fort Pelly on the Swan River.

Highway 15 continues east past Melville and Esterhazy towards

Fort Ellice. North of Spy Hill, the Carlton Trail once again turned east where a sign marks its location. This segment of the trail is called Ellice Trail and continues on to St. Lazare which is actually the site of the original Fort Ellice built on the west bank of the Assiniboine River.

From St. Lazare, the Carlton Trail struck southeast along what is now Highway 4, passing Shoal Lake and finally arriving at the Red River settlement.

Hay Lakes–Battleford Telegraph Trail

On a grassy knoll near the village of Hay Lakes, cattle graze where once the clicking messages (and occasional checker games) of Canada's most westerly telegraph station were heard. The alien sounds were first heard over a hundred years ago, at a time when proud natives hunted the buffalo, and the telephone was yet to arouse Western Canada from its rural slumber.

In 1875, the Canadian Government began construction of what was to be a transcontinental telegraph line from Thunder Bay to the Pacific Ocean. The route was by way of the Yellowhead Pass, the provisional right-of-way of the Canadian Pacific Railway. It was believed that the telegraph would assist both in the construction of the railroad and in governing the vast western frontier. In 1874 the North West Mounted Police made their epic trek across the southern plains of what is now Manitoba, Saskatchewan and Alberta. Soon they established posts at Fort Macleod (1874), Calgary (1875) and Fort Saskatchewan (1875). In order to expedite routine police business, a telegraph line was necessary.

In 1874, contracts were entered into to build various sections of the telegraph line from the Great Lakes to connect with the Telegraph System of British Columbia. In 1874, a contract with Sifton, Glass and Fleming, of Winnipeg, was made to construct a telegraph line and clear a path 132 feet wide from Selkirk, Manitoba, to Livingston, or Swan River.

Richard Fuller was awarded the contract to construct the section of the telegraph extending from Fort Pelly to a point just south of Fort Edmonton (near present-day Leduc). Construction began in June of 1875. Led by George Wright, a supply party was making good time until it reached Fort Carlton. There two Cree chiefs, Star Blanket and Big Child, inquired of Wright just what he was doing with the long train of carts loaded with exotic-looking insulators and alien wire that would scare away the game.

Citing a proposed treaty, the chiefs allowed the supplies to be laid along the line, but warned Wright against the further destruction of Cree land. Wright went to great lengths to solve the problem, but to no avail. Three scouts sent by Big Child later stopped the supply party just west of the Four Blackfoot Hills near Grizzly Bear Coulee. One of the scouts voiced these prophetic words:

> My chief is Big Child. He told me to tell you not to go any further until a treaty is made. . . . We want the government to send a man to tell us what their intentions are, whether the government likes us or hates us. It is going to make a treaty with us or it is going to take our country by force. When I saw the wire. . . I felt as if something was sticking in my throat. We are only three scouts, but if you insist on going ahead we will soon get enough Crees to make you stop. All our people are of the same opinion. When a treaty is made, our rights to the land and country will be the government's. Until then, the rights are ours, and we tell you to stop!

Wright honoured the Crees' wishes and later in 1876 when the treaty was made, Star Blanket and Big Child were the first to sign. (Big Child was later implicated in the killing of a Fort Walsh North West Mounted Policeman.) It was Alex Taylor who completed what to Richard Fuller was a cursed project. In 1877, Taylor brought the line as far west as Hay Lakes, only to discover that no one lived there. Assuming that it takes at least two to communicate — or one to listen — he grounded the cable in Hay Lakes and dispatched a message to Ottawa, informing his superiors that, contrary to his instructions, Hay Lakes was not peopled. Ottawa's reply:

> A pox upon you for your impertinence, sir. That godlike individual, the Honourable Minister of the Interior, has directed that there is a settlement at Hay Lakes. Therefore, you will open a telegraph office there, as directed.

Lest he lose his contract — or worse, his pension — Taylor complied. But being a man of practical vision, he was resolved to handle the problem in his own way. He judged that if the Ottawa gang was that far removed from the situation, they probably wouldn't know the difference if he extended the line to Edmonton anyway. Needless to say, Edmonton's merchants were quite pleased with the fruits of progress being laid before them — and for once without even having to ask. Taylor had lots of cable and insulators, but no poles left, so a crew of Edmonton volunteers set about felling trees, little considering the threat of brush fires. (From Battleford to Hay Lakes, the telegraph poles were made of iron.)

Summer nights could be a nightmare as the prairie horizon glowed with fire started by lightening. The telegraph line was in constant peril. Even the most trivial of happenings — windstorms or floods — left it inoperative.

And yet, despite these interruptions, the telegraph served the Edmonton community well. Alex Taylor was a respected resident for over 40 years. Over the wire, Taylor scooped the hot news stories and weather reports that he and co-founder Frank Oliver printed in the *Edmonton Bulletin*.

For a while the telegraph was a useful device, but what is said of the "best laid plans of mice and men" certainly rang true with the telegraph line. When the Canadian Pacific Railway tracks were ultimately laid between Regina and Calgary, instead of along the painstakingly built telegraph line, settlement paced with the iron horse. Along the railway, towns sprang from the dust like the false fronts of a prairie festival, but along the Hay Lakes–Battleford Telegraph Trail, only a handful managed to survive. Not wanting to kill its white elephant, the Canadian government kept the telegraph line to Hay Lakes open for official business. With a scarcity of doctors, medical advice was transmitted free of charge, and the operators played their occasional long-distance checker games. More importantly, the telegraph was instrumental in helping to police the unruly west, notably the Riel Rebellion of 1885. However, in 1887, the telegraph line was re-routed from Battleford to Edmonton along the Carlton Trail via Fort Pitt and Victoria.

The Trail to Hay Lakes

Leaving Selkirk, Manitoba, the telegraph trail stretched in a northwesterly direction, crossing the narrows of lakes Dauphin and Winnipegosis to Livingstone, a trading post on the Swan River near the Manitoba–Saskatchewan border. Departing Livingstone, the line veered west by north, skirted Humboldt, and crossed the Carlton Trail. It closely paralleled the present railway from Humboldt to Langham and continued along the north bank of the North Saskatchewan River to Battleford. The line continued through Battleford to Eleanor, Saskatchewan, where a telegraph station was built. The telegraph trail then struck westward between the hamlet of Round Hill and the north shore of Lake Demay, a few miles south of Kingman. Passing just north of Armena, the trail originally ended at Hay Lakes, until 1879, when it was extended to Edmonton via Leduc along the Edmonton–Calgary trail.

By most accounts, the Hay Lakes–Battleford Telegraph Trail should be called the Hay Lakes Trail, as it existed and was well travelled before the telegraph line was constructed. Though in places the trail didn't follow the exact route of the telegraph line, it never diverged by more than a mile. As early as the 1870s much of its track was well defined by wagon ruts, in places two feet deep. Countless Red River carts carried freight and settlers west along both it and the Carlton Trail from which it branched.

In the decade before the turn of this century, the first pioneers with their cattle and pigs straggled in to settle the area between New Sarepta and Hay Lakes. At that time the country was so wet

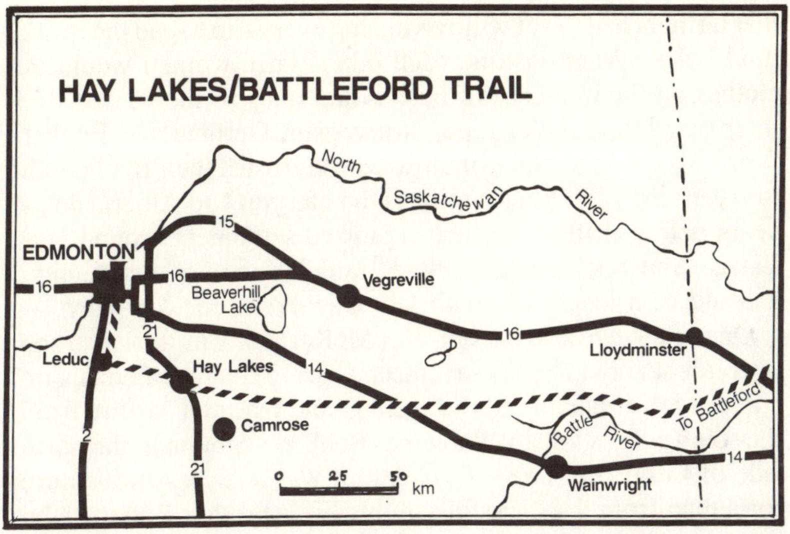
HAY LAKES/BATTLEFORD TRAIL
North Saskatchewan River
EDMONTON
16
15
16
Vegreville
Beaverhill Lake
21
Leduc
Hay Lakes
14
16
Lloydminster
To Battleford
Camrose
2
21
Battle River
0 25 50 km
Wainwright
14

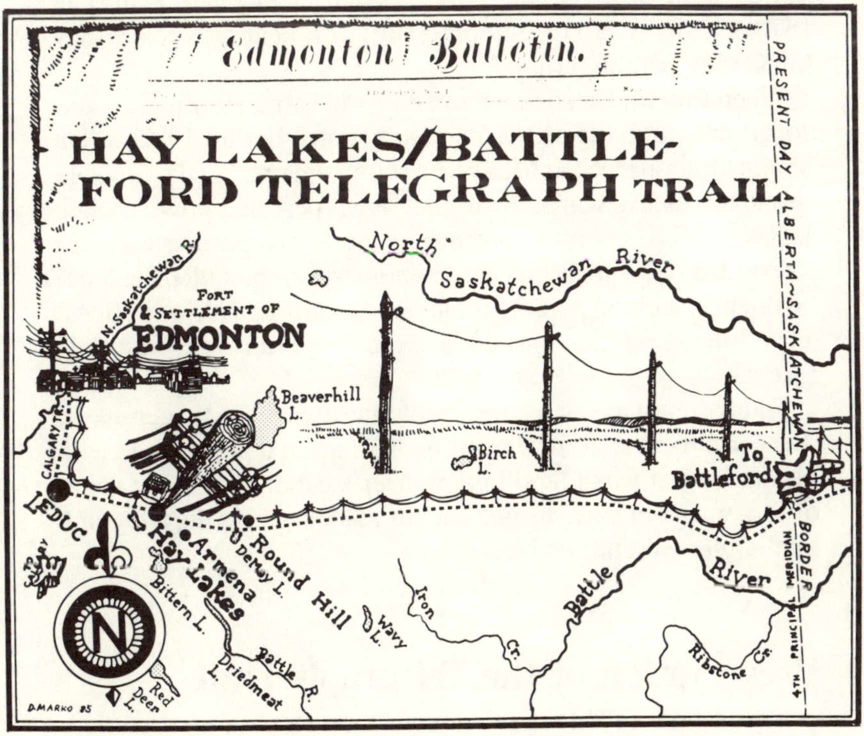
Edmonton Bulletin.
HAY LAKES/BATTLE-FORD TELEGRAPH TRAIL
PRESENT DAY ALBERTA~SASKATCHEWAN BORDER
N. Saskatchewan R.
FORT & SETTLEMENT OF EDMONTON
North Saskatchewan River
Beaverhill L.
Birch L.
To Battleford
CALGARY TR.
LEDUC
Hay Lakes
Armena
Round Hill
DeMay L.
Wavy L.
Iron Cr.
Battle River
4TH PRINCIPAL MERIDIAN
Ribstone Cr.
Bittern L.
Driedmeat L.
Battle R.
Red Deer L.
N
D.MARKO 85

that farmers had to cut willows during every trip to pad the trail's mud holes. When visiting each other, farm women would tie clothes on the branches of bushes and trees so they could find their way home. Over time, Norwegian, German and English farmers joined the ranks of those who entrusted their livelihoods to Alberta's soil. Other families who had come to Alberta a few years before drifted into the organized settlement around Hay Lakes. And yet, despite this wide-ranging sense of community, it could be a lonely life with few neighbours for company.

One of the earliest settlers, Jim McKernan, was the telegraph operator at Hay Lake for seventeen years. To supplement his income, McKernan turned to ranching and fenced a pasture from the south side of what is known as Bakken's Slough to the north side of Little Hay Lake. Evidently it was a successful venture for a man from a successful family. Jim's brother Robert (who helped build the telegraph and who lived at Hay Lakes in the late 1870s) settled south of Edmonton in 1879, near the present-day McKernan neighbourhood.

When Bob McKenzie took over the McKernan farm, the abandoned telegraph office made a dandy barn, but by 1902, when Simon Grue succeeded McKenzie, little remained of the tired old shack but its log walls. Even they have perished now. As late as 1902, a considerable portion of the telegraph line was still suspended. It soon disappeared as homesteaders settled the land, no doubt snatching poles to mend fences or prop up clotheslines. That thin wire which once etched the blue prairie sky vanished forever.

Today, remnants of the telegraph line are valuable keepsakes. A second-hand store in Vermilion has part of a poplar pole with insulators that was found in a pioneer's cabin. Also, just south of Ryley, an old hand-forged anchor that once held the telegraph line is marked and visible.

Preservation of the Telegraph Trail

Mel Helmer of Viking is a devotee of historic trail discovery.

TELEGRAPHIC.

WINNIPEG, July 14, 1885.

No important foreign despatches.

Big Bear and several others are now on the way to Regina from Prince Albert.

Scouts have traced Little Poplar and band from near Battleford, and report him on the way to the boundary.

Gen. Brackenbury telegraphs from Fatniah, Egypt, that a letter which was received there on the 10th inst. states El Mahdi is dead.

No word has yet been received from the troops since they left Prince Albert. They are expected to reach Selkirk to-day or to-morrow.

Riel's counsel have arrived. They are F. X. Lemieux, M.P.P., Charles Fitzpatrick and J. W. Greenshields. All will appear at the opening of the trial on the 20th.

Poundmaker and twenty-two other prisoners from Battleford have arrived at Regina. Of this number ten men were sentenced at Battleford and are being brought to Stoney mountain.

Orange and green riots took place at Waterford, Belfast and a few places in England on Sunday, 12th inst. At Waterford several civilians were stabbed to death by soldiers. The populace became furious and many soldiers were maltreated.

Excerpt from The Edmonton Bulletin, *18 July 1885. The telegraph was Edmonton's primary source of news from the outside.*

Though he and others have gone to great lengths to preserve the Hay Lakes–Battleford Trail, they have to date received little support. Trekking and staking miles of trail is not easy work, but it has its rewards. Unfortunately, preserved trail marks too often fall prey to senseless vandalism or theft.

Just down the road from the Helmer place lived the parents of Klondike poet Robert Service. They are now buried in the Mannville cemetary. A bit further west, a detachment of North West Mounted Police operated a telegraph station, of which only rock and a cellar remain. The telegraph was instrumental in helping to quell the Riel Rebellion of 1885.

Two miles west and one mile south of Holden, Bruce and Walter Loutz have erected a monument and a bronze plaque that identifies the trail and conveys a bit of its history. A sign placed near the Round Hill Hotel identifies the trail extending just south of the east-west road running through town.

The Hay Lakes–Battleford Telegraph Trail passed within 100 feet of the telegraph station now marked by a cairn at Hay Lakes. Raymond Stavne, who married one of Simon Grue's daughters, lives on the Grue homestead. The trail is preserved on the Stavne's property, and will soon be developed as an historic site complete with an interpretive facility.

Other remnants of the Hay Lakes Trail lie within the Third Prairie Steppe beginning at Alberta's eastern border. From its Leduc terminus, proceeding eastward, it passes through an area of level to gently rolling topography just west of the Cooking Lake moraine area. Most of this area has been subjected to post-glacial wind and water deposition, so the slopes of the hills are long and smooth.

The Cooking Lake moraine is a hilly area of knobs and kettles between Tofield to the east and Cooking Lake to the west. The Hay Lakes-Battleford Trail passes along the southern edge of this area, and then strikes eastward through the undulating expanse and shallow depressions that lie west of modern-day Highway 36.

To the east, the Viking moraine is a hilly upland area which contains many sloughs, lakes, deep coulees and river channels,

DAVID CUNNINGHAM

Abandoned farmstead near Bruce, Alberta. The Telegraph Trail passed just south of Bruce, and crossed modern-day Highway 14 at Viking.

of both pre- and post-glacial origin.

Further east, the trail passes through the Vermilion bedrock zone characterized by gentle slopes of hills without any sharp breaks. A thin veneer of drift fails to mask the irregularities of the underlying sandstone bedrock which in spots has been contorted by the pressure of overriding glaciers.

The Lea Park Lowland, an area on the eastern border of the sandstone bedrock expression which is level to undulating, extends from St. Paul, Alberta,1 to Lloydminster. The Telegraph Trail passes near the Blackfoot Hills which are located about 15 miles southwest of Lloydminster. The hills are prominent, rising some 300 feet above the surrounding area.

The Whoop-Up Trail

The Fort Benton–Whoop-Up Trail has a colourful — though nefarious — history. For over 200 years this trail was a warpath of slaughter, a road to riches and a way of sorrows — a well-worn path to Alberta's rich prairie pasture, teeming with game and caches of gold, but the dwindling domain of the Blackfoot, Blood and Peigan Indians.

The Whoop-Up Trail brought the lawless edge of the white world to bear on the desperately dwindling native tribes. (The smallpox epidemic of 1869-70 killed as many as half the population of the Plains Indians.) For decades, the Blackfeet Confederacy had ruled over the plains of southern Alberta, preserving the area against the encroachment of the pioneers. Unlike the natives of the north, who were systematically organized by the Hudson's Bay Company to reap the furry benefits of its monopoly, the powerful Plains Indians organized themselves and traded on their own initiative with either the Hudson's Bay Company to the north or the American fur company to the south. With whiskey for a weapon, the first permanent "settlers" in southern Alberta were at once ruthless and brave, willing to take on a country none but the Blackfoot dared to traverse, but unwilling to give any quarter to the warriors of the plains. No substantial settlement would occur until the arrival of the North West Mounted Police in 1874.

PROVINCIAL ARCHIVES OF ALBERTA

A mule train is loaded up and ready to leave Fort Benton, Montana, headed for The Whoop-Up Trail (1877).

Also known as the Fort Benton Trail and the Old North Trail, the Whoop-Up Trail extended from Fort Benton, Montana, to Fort Whoop-Up at the confluence of the St. Mary and Old Man rivers, southwest of present-day Lethbridge. Fort Benton, the "Chicago of the West" at the head of navigation on the Missouri River, was an important center of commerce. It was served by sternwheelers from St. Louis and other central American cities.

The Whoop-Up Trail was a segment of the Old North Trail, an Indian trail that spanned the length of North America, paralleling the Rocky Mountains. Perhaps as long as 30,000 years ago, nomads trekked the Old North Trail during the great migration from ancient Asia, across the Bering Strait (then a land bridge) and southward along the Continental Divide.

The Whoop-Up Trail can still generally be traced from Calgary to Fort Macleod along Highway 2, crossing the Blood Indian Reserve west of Lethbridge and continuing south along Highway

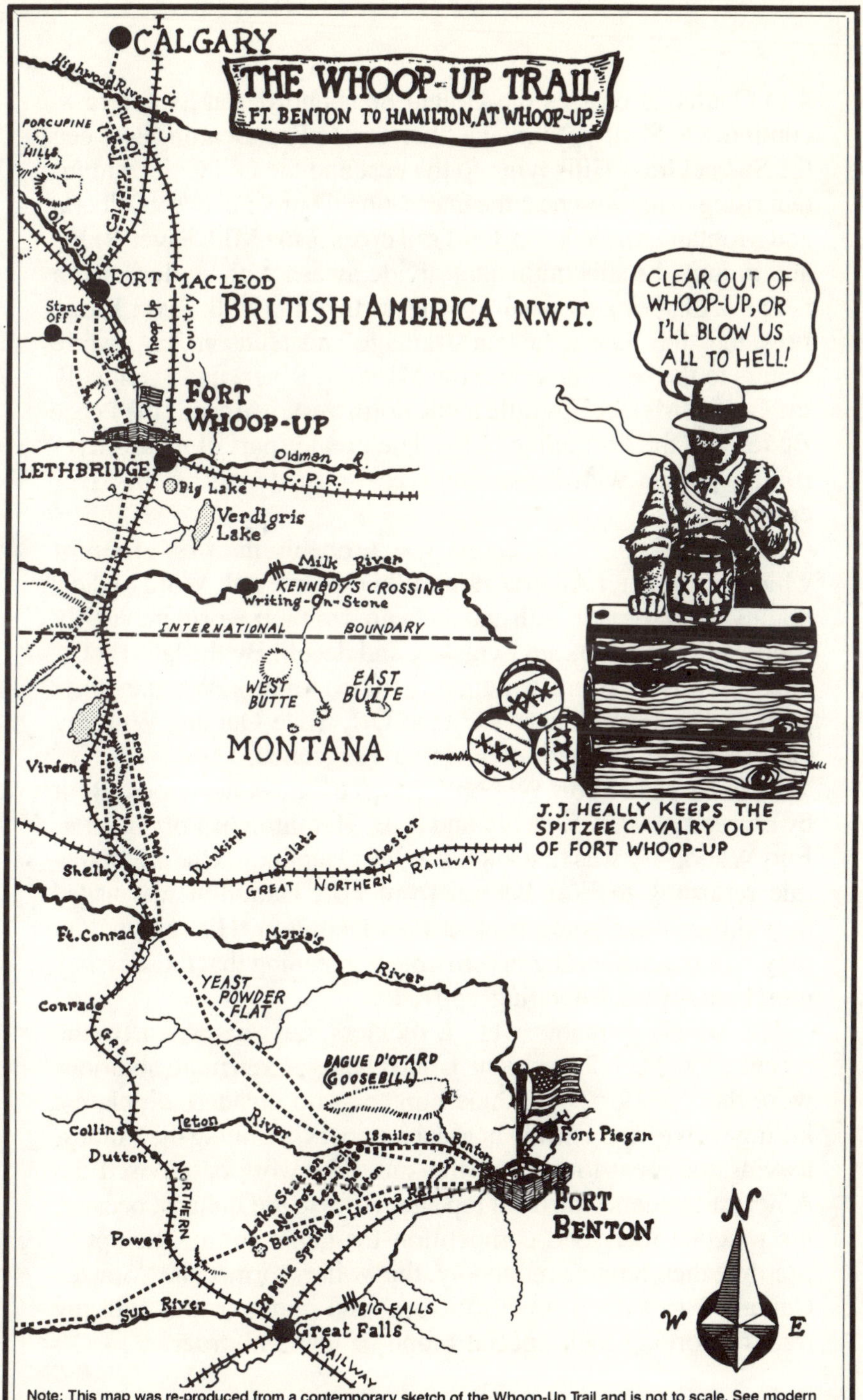

Note: This map was re-produced from a contemporary sketch of the Whoop-Up Trail and is not to scale. See modern highway map for true porportions and distances.

4 to Coutts. It crosses the United States border at Sweetgrass, continues to Shelby, Montana, then east to Fort Benton. Between the Sweet Grass Hills lying to the east and the Old Chief Mountain rising in the west near the international border between Alberta and Montana, the Whoop-Up Trail crosses the Milk River Ridge and mounts its subcontinental divide awash with coulees and a network of springs. (Waters to the north of the Milk River Ridge flow into the Saskatchewan drainage and Hudson Bay, while waters to the south flow into the Missouri River, and eventually the Gulf of Mexico.) A little to the northwest, it ascends the Peace Buttes and the Porcupine Hills. The greater part of it runs over rolling prairie with brush and trees located only at the river crossings.

A party of gold prospectors was probably the first group of white men to enter Alberta along Whoop-Up Trail. Word of their bounty soon spread south and the wares of their world inevitably followed: guns, tobacco, whiskey and deceit. By the late 1860s, whiskey was the prime commodity traded for furs. Whiskey forts with such colorful names as Stand Off, Slide-Out and Whiskey Gap reeled to life with consuming vengeance.

Of these posts, Fort Whoop-Up was the most notorious. Built by entrepreneurs J.J. Healy and A.B. Hamilton of Fort Benton, Fort Whoop-Up was first known as Fort Hamilton. One day someone returning to Fort Benton from Fort Hamilton was asked how things were going back at Fort Hamilton. His reply: "Oh they're still a whooping of 'em up" — meaning that the whiskey merchants were whooping-up trade.

The whiskey traders were a reckless sort, hardened by the recently fought civil war in the United States. Even more notorious were the "wolfers" and hide-hunters who invaded the plains, killing wolves and buffalo in great numbers, skinning their hides, leaving the meat to spoil in the sun. The wolfers opposed the American traders who dealt repeater rifles to the Indians, because the practice increased competition for game. In an attempt to preserve their hunting monopoly, the wolfers formed the "Spitzee Calvary" to swoop down on forts and confront the whiskey traders, forcing their special brand of law and order.

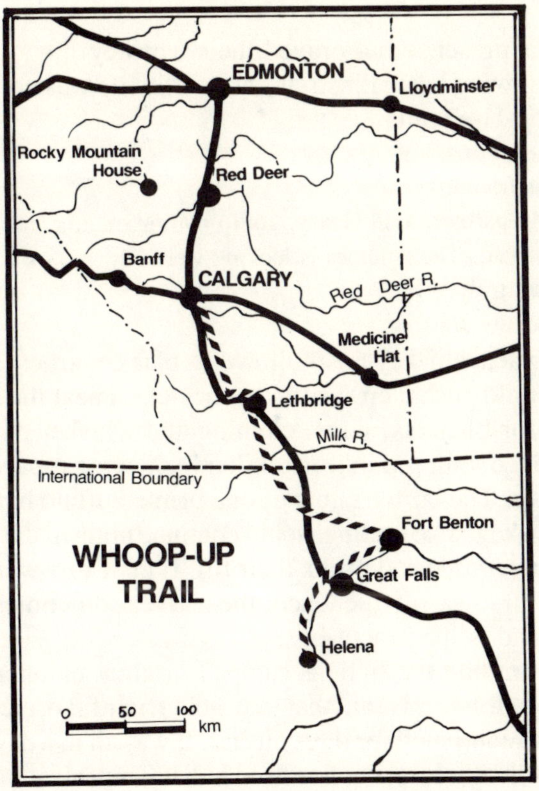

The Spitzee Calvary once surprised J.J. Healy at Fort Whoop-Up. While they threatened the fort with swift violence, the quick-thinking Healy casually heated a fireplace poker, strolled over to a keg of gunpower and promised to blow the whole place to hell if they didn't leave. They retreated. Healy, a man of many interests, was later the sheriff of Montana's Chouteau County and the mayor of a Yukon shantytown during the Klondike Gold Rush.

In every way these lawless men spoiled the land. The wolfers with their strychnine-laced meat lured and killed wolves. The mercenary hide-hunters, in their quest to supply leather for belts to drive the insatiable industries of the eastern seaboard, denuded the plains of buffalo, leaving behind rotting carcasses. The whiskey merchants, with their narcotic liquor, debauched the Indians who hated their presence but craved their whiskey.

The following letter illustrates how little life was valued along the Whoop-Up Trail:

Whoop-Up

> Dear friend:
>
> My partner, Will Geary, got to putting on airs and I shot him dead. The potatoes is looking well.
>
> Yours truly,
>
> Skookum Jim

Trading at a whiskey post followed a black-market ritual. The Indians would push their furs through wickets near the main gate in return for blankets, guns, clothing and whiskey. The liquor was seasoned with red ink, painkillers, chewing tobacco and/or castile soap. The barter rate was one prime buffalo hide for one cup of whiskey. The trading would continue until all the furs were traded or everyone had drunk their fill. If their furs were all sold but their thirst wasn't quenched, the natives sometimes stormed the forts and helped themselves.

During trading the Indians camped in great numbers around the whiskey forts. It is said that they held frightful orgies and war dances that continued for days on end. It was all in a day's march to find the bloody corpses of scalped Indians and white men strewn along the Whoop-Up Trail.

The flagrant excesses of the whiskey trade and the infiltration of unscrupulous American interlopers culminated with the Cypress Hills Indian massacre in 1873. White wolfers from Fort Benton, allegedly searching the Cypress Hills for a band of Indians accused of stealing some horses, attacked a small encampment of innocent Assiniboines. At least 30 were killed. Western lawlessness had become so rife that Ottawa was compelled to react — to curb the violence and protect Canadian interests against the "manifest" American expansion. The government formed a semi-military organization, the North West Mounted Police, and dispatched troops to strategic areas throughout Alberta in an effort to curb illegal trading and assert the sovereignty of the young Dominion. By orders of Colonel French, one division was sent to Fort Edmonton and another to Swan River. A third, led by Colonel James

ALBERTA CULTURE

Looking west from the remains of Fort Whoop-Up. The slight depression in the foreground marks a former wall. (From Alberta Culture's report, An Inventory and Historical Description of Whiskey Posts in Southern Alberta, *by M.A. Kennedy and B.O.K. Reeves.)*

Macleod, came by way of Fort Benton and the Whoop-Up Trail into the middle of the whiskey-trader's domain.

Macleod and company assumed that the source of trouble, Fort Whoop-Up, would be the logical place to establish their headquarters. However, they arrived only to find the place in shambles and one old man, David Akers, left in charge, tending his garden. Healy offered to sell the fort to MacLeod for the outrageous sum of $20,000 U. S. dollars but was flatly turned down. The notorious fort, having seen the last of its glory, was closed for good.

South of Lethbridge along Highway Number 5, a sign commemorates the notorious fort. The sign stands near where the Whoop-Up Trail once passed. A replica of Fort Whoop-Up in Indian Battle Park in Lethbridge displays much of the fort's original structure and pioneer furnishings.

Being unfamiliar with Alberta's turf, and in need of a guide, the police hired Jerry Potts, son of a Scots father and Peigan mother. He led them to an island on the Old Man River where they established their headquarters, Fort Macleod, the first police fort built in the west. Potts served the force as an interpreter and scout for many years until his death in 1896. He was greatly respected by the Blackfoot tribe for his bravery in battle and fair-minded dealings. Along Highway 4 between Sterling and Warner, an historic plaque honors Potts as being "short," "bow-legged," and a "credit" to the force.

Once settled at Fort Macleod, the force was confronted with the merciless task of having to police some thousands of square miles of virgin territory. Where troubled brewed, the police arrived posthaste, building forts and blazing trails to meet their needs. Whoop-Up Trail — no longer "whooping" — was further extended in a network of roads linking forts Macleod, Calgary (established in 1875) and Walsh in the Cypress Hills. Although there were probably no more than twenty settlers in what is now southern Alberta, a community developed around Fort Macleod, including a livery stable, restaurants, shops, a pool hall and even a church.

The original fort and town were relocated in 1883 atop the bank of the Old Man River. Nothing of the original fort was preserved and even its traces were washed away by the river tide. Every now and then, however, a square-head nail is found, a treasured reminder of the pioneer days. The North West Mounted Police Museum of Canada at Fort Macleod displays an exact reproduction of old Fort Macleod, complete with artifacts from the whiskey trade era, including armaments, personal effects of the police troops and Indians, and replicated buildings of the original town.

With the police on guard, the whiskey trade was put to death. The wolfers and hunters, having chased the buffalo to the brink of extinction, left for parts unknown. Favorable circumstances led the government to promote southern Alberta as a peaceful area open to settlement. In 1876, the Old North Trail was bonded for freighting. It yielded a quicker and cheaper transport of goods than did the Winnipeg route. In its boom years between 1874 and 1885, the Old North Trail carried one-third of all freight handled through Fort Benton. Teams of oxen, mules and, later, horses hauled supplies for the N.W.M.P., novice ranchers, and storekeepers at Forts Macleod, Calgary and other posts in southern Alberta.

Each year, come winter's thaw, the wagon trains began their exodus from Fort Benton, plodding along at a funeral's pace, covering perhaps 12 to 15 miles a day. For greater safety, these "bull trains" were organized into as many wagons as could be physically managed. As many as eight teams-each formed of 12 to 15 yoke of oxen-pulled wagons hitched in tandems of three and carrying freight exceeding ten tons.

In 1883, the I.G. Baker Company — never one to miss an opportunity — began a bi-monthly stagecoach service between Fort Benton and Fort Macleod, making the trip of over 200 miles in five to eight days. Later in 1890, when the railway between Calgary and Edmonton supplanted its service, the I.G. Baker Company sold its Canadian interests to the Hudson's Bay Company. The last ox-drawn wagon haul from Fort Benton to Fort MacLeod occurred in 1892. The trail still functioned for many years, however, as an important settlement route travelled by immigrants to Canada from the United States. The Whoop-Up Trail was surveyed in 1894, and can generally be traced along Highway 2.

Portions of the Whoop-Up Trail still exist in their natural state.

On the Canadian side of the international border, the trail is visible for about a quarter of a mile. An historic sign along Highway 2 near the border indicates its nearby location.

At the watershed of the Milk River Ridge near the present-day town of Milk River, though broken by fences, the trail extends through pasture for about two miles.

In 1964, a group of boy scouts marked about six miles of trail in Canada along the west side of Highway 2.

At the Federal Government Interpretive Site at Fort Whoop-Up, wagon ruts are evident across the valley, as well as at the Pinhorn Provincial Grazing Reserve near Kennedy Crossing.

These remnants of the Whoop-Up Trail are testimony to the volume of traffic that passed over the barren prairie. With the colourful legacy of places (Whoop-Up, Standoff and Slide-Out) and names ("Liver-Eating" Johnson, "Dutch" Wachter and "Kamoose" Taylor) , there is a roguish nostalgia to the era of the Whoop-Up Trail. The trail corrupted a people, making some men rich and others dead. Yet it was also the catalyst that prompted the establishment of a police force to introduce law and order, which in turn finally opened the prairies for settlement. This was the west at its wildest.

The Calgary–Edmonton Trail

Departing Edmonton, going south, it's the "Calgary Trail," but from Calgary to Edmonton it's the "Edmonton Trail." A matter of perspective.

In 1875, Fort Edmonton was a respectable and growing community, a long-standing bastion of the Hudson's Bay Company fur-trading empire. Calgary was little more than a patch in the mud. It was Inspector E. A. Brisebois of the North West Mounted Police's "F" Troop who had founded "Cow Town" at the confluence of the Elbow and the Bow Rivers. A fort was built in 1875 and unofficially christened "Brisebois," but the name was short-lived. (Brisebois himself was eventually found unfit to command and mustered out of the force.) Colonel James Macleod, the man responsible for the N.W.M.P. in southern Alberta, suggested that the fort be named "Calgary" after a family estate on the Island of Mull in Scotland. Assistant Commissioner A.G. Irvine accepted Macleod's recommendation. Macleod claimed the name meant "clear running water" in Gaelic, but subsequent scholarly research has proven Macleod wrong, and a definitive meaning is still lacking.

For over a century, the forerunner of the Calgary–Edmonton Trail was an important conduit of communication and movement between the open prairie and Fort Edmonton. Situated in the vast expanse of a glacial corridor, this trail — a segment of the Old

North Trail — was a natural pathway. To its east, dense forests; to the west, gently rolling hills. The trail ebbed and flowed with the cropped grasses of the parkland fringe. It crossed shallow dales and gentle fords of the Red Deer and Battle rivers, meandering through the pre-glacial Red Deer Valley.

Indian trails running south from Edmonton had long been adapted and used sporadically by early explorers and itinerant traders. Alexander Henry and David Thompson recorded their travels over Wolf Track, a segment of the trail striking from Edmonton to Rocky Mountain House. (See The Rocky Mountain House Trail.) In 1858, the Palliser Expedition camped at an area then known as "Cache Camp" — later called Poplar Grove and today known as Innisfail.

However, it wasn't until 1873 that a "civilized" trail between Calgary and Edmonton came into being when the Reverend John McDougall, a Methodist missionary, and his brother David, a free trader, built a mission at Morley. They cut a formidable path for their cart train and livestock from Fort Edmonton through the Peace and Bear hills to the crossing at Red Deer and then on to the Lone Pine stopping house thirty miles south. From Lone Pine (near present-day Bowden) they veered westward up the Little Red Deer River Valley and then down the Ghost River to the Morley site on the Bow River, some fifty miles upstream from Calgary. When Fort Calgary was built two years later, the trail from Lone Pine was extended due south along what is now modern Highway 2.

The Calgary–Edmonton Trail became a vital extension to the expanding marketplace of the American traders during the 1870s and 1880s. Once the North West Mounted Police eradicated the whiskey forts, of course the whiskey traffic didn't stop, but continued clandestinely in barrels labelled "Coal Oil." Intensive settlement began in what is now southern Alberta. Though the Calgary–Edmonton Trail was originally blazed from Lone Pine to Calgary by the Hudson's Bay Company, the I.G. Baker Company and others from Fort Benton, Montana, soon dominated the burgeoning pioneer market. The Hudson's Bay Company and other Canadian traders were forced to transport goods from

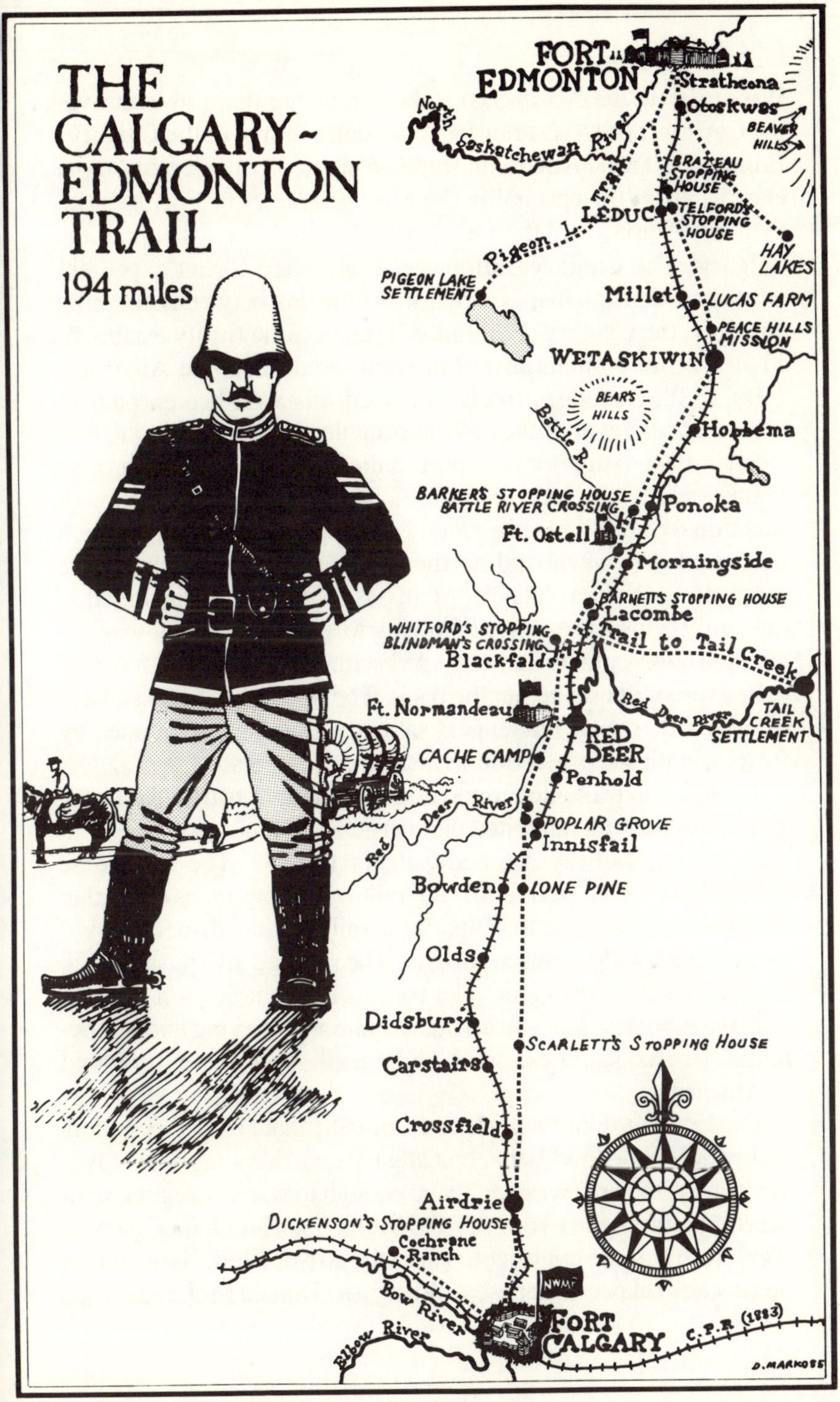
THE CALGARY-EDMONTON TRAIL
194 miles
FORT EDMONTON
Strathcona
Otoskwas
BEAVER HILLS
North Saskatchewan River
BRAZEAU STOPPING HOUSE
LEDUC
TELFORD'S STOPPING HOUSE
HAY LAKES
Pigeon L. Trail
PIGEON LAKE SETTLEMENT
Millet
LUCAS FARM
PEACE HILLS MISSION
WETASKIWIN
BEAR'S HILLS
Battle R.
Hobbema
BARKER'S STOPPING HOUSE
BATTLE RIVER CROSSING
Ponoka
Ft. Ostell
Morningside
BARNETT'S STOPPING HOUSE
Lacombe
WHITFORD'S STOPPING
BLINDMAN'S CROSSING
Trail to Tail Creek
Blackfalds
TAIL CREEK SETTLEMENT
Red Deer River
Ft. Normandeau
RED DEER
CACHE CAMP
Penhold
Red Deer River
POPLAR GROVE
Innisfail
Bowden
LONE PINE
Olds
Didsbury
SCARLETT'S STOPPING HOUSE
Carstairs
Crossfield
Airdrie
DICKENSON'S STOPPING HOUSE
Cochrane Ranch
NWMP
Bow River
Elbow River
FORT CALGARY
C.P.R. (1883)
D. MARKO 85

Winnipeg up the North Saskatchewan or lug them by cart over the Carlton Trail to Edmonton, and then south over the Calgary–Edmonton Trail. American trades freighted considerably more cheaply from the east along the Missouri River to Fort Benton, then by wagon over the Old North Trail to Fort Macleod and Calgary. The Calgary–Edmonton Trail made for an expedient northern extension to this network. As the link between economic domains, the Calgary–Edmonton Trail became firmly established as the pre-eminent artery in what would become Alberta.

By 1888, travellers freely ventured this well-beaten path in whatever conveyance they found suitable. Among these, the Red River cart was superior transport, quite adaptable for use on varying terrain. It was soon discovered that heavier wagons, though functional on solid prairie ground, were poorly suited to the rich black soils of central and northern Alberta. Stagecoaches were more "stage" than "coach." Patrons soon grew accustomed to the unnverving experience of having to sit in the back of wagons covered with canvas and crowded with freight. A sudden stop called for swift escape out the back. The open coaches used later offered little relief. Passengers sat in the middle, surrounded by freight, with no protection against the elements.

Though the trail accommodated commercial traffic for many years, few settlers inhabited the surrounding land until the Canadian Pacific Railway reached Calgary in 1883. The iron horse ushered in a boom period for the trail to Edmonton, as Canadian supplies were shipped to Calgary via rail and then dispersed over this sole link with northern Alberta. The railroad also pushed back the American interlopers. The trail to Edmonton soon became one of the busiest roads in all the Dominion, carrying about nine-tenths of the supplies, mail and people that were freighted northward.

The *Edmonton Bulletin* reported in 1881 that about 363 settlers had peopled the land between Calgary and Edmonton. Just two years later, the area was populated enough to warrant regular mail service between the two cities. Ad McPherson and John Coleman were awarded the contract to carry the Royal Mail, light freight and occasional passengers. Later that year, Donald McLeod started

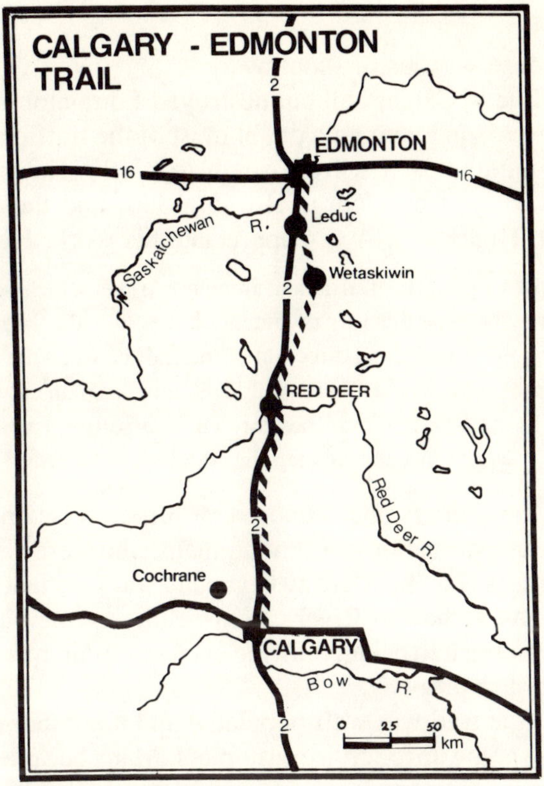

the first weekly stagecoach service. The stage would leave Jasper House, Edmonton, at 9:00 a.m., and the steamboat dock at 9:30 every Monday, making stops at Wetaskiwin (then Peace Hills), Ponoka (then Battle River), Red Deer Crossing and Willow Creek (in the Olds area), arriving in Calgary the following Friday if all went well.

Freight wagons took considerably longer — up to 14 days, depending on the weather conditions. The trail was often mired and deeply rutted. Frequent flooding and blizzards stalled travel. The *Edmonton Bulletin* recounted the ordeal of one poor stagecoach driver who, while bound for Calgary, lost his way in a freezing storm. He unhitched the horses, tied them to the sleigh, and spent the whole chilling night walking around his outfit to ward off death. When the rising sun brought visibility, he looked around and came to realize that his tortuous night was spent in vain. The Lone Pine stopping house was just a shout away.

The rail line to Calgary all but destroyed Edmonton's role as a supply centre, and without a doubt most of the traffic between the two communities was from south to north. The trail was officially surveyed by George Roy in 1886, and the modern traveller on Highway 2-A can appreciate his work. He wrote:

> In view of the great traffic and immence travel which some day may be done this way my intention was to make the road as straight as the actual direction of the trail between the two extreme points would allow. A little ditching, a small culvert, a slight cut or a few branches thrown on a soft spot will be all that will be needed to keep the trail open.

In 1891, the Calgary and Edmonton Railroad was completed, much to the consternation of Edmontonians who were distressed that the Canadian Pacific Railway Company was reluctant to cross the North Saskatchewan River. As a result, the new town of Strathcona sprang into being, drawing even more commerce away from Edmonton proper.

By 1885, the trail was well-populated and a number of stopping houses sprang up to service customers. Many became sidings of the Calgary–Edmonton railway, and later evolved into full-fledged towns and cities.

Calgary to Lone Pine

A cairn in Calgary's Central Park and a store near the Canadian National freight sheds commemorates the early police fort at the site of modern Calgary. The store at the Canadian National freight sheds is situated on the original site of the North West Mounted Police reserve.

One possible stopping house on the Calgary–Edmonton Trail was the Wessex house. As late as 1963 a stone pile, remains of the former stopping house, could be seen in a pasture adjacent to Old Coach Road, an access north of Wessex on old Highway 2 about one-half mile east of the highway.

Hudson's Bay Company carts, loaded with furs from the north, queue up to load a train in Calgary, 1888. With the arrival of the railroad, traffic to and from the north increased on The Calgary-Edmonton Trail.

PROVINCIAL ARCHIVES OF ALBERTA

Near Rosebud Creek — sometimes referred to as Serviceberry Creek or Savasse Berry Creek — Sam Scarlett operated a mail station. A nearby North West Mounted Police outpost was established in 1886 for the purposes of escorting Scarlett on his rounds. Located approximately two miles east and three and one-half miles north of Carstairs, the original frame dwelling — a long narrow building with connecting rooms — was constructed of lumber freighted across country from Medicine Hat before 1885. Sam Scarlett was a coal miner on the banks of the Rosebud Creek. A large rock near the site at Willow Creek Coulee is engraved with the names of early travellers. Sam Scarlett's house was undoubtedly one of the earliest along the Calgary–Edmonton Trail.

Lone Pine to Red Deer Crossing

One of the more popular stopping houses, run by Jean Langlois, Lone Pine was an ancient Indian meeting place, conveniently located at the confluence of the Blackfoot Crossing Trail and the branch roads leading to Morley and Rocky Mountain House. Many travellers camped at this area, just east of present-day Bowden, before stopping houses afforded more comfortable arrangements. The Blackfoot name for Lone Pine—quite a mouthful—is *Minnie-Hay-Gwack-Pack-Waksut*. Lone Pine house later served as a local mail station. Near Lone Pine, a man named Batho owned a remodeled cookhouse which he had purchased from the railway. He operated the Lone Pine house after Langlois was forced to move his operation when the railway bypassed his place.

Just north of Lone Pine and east of Bowden, James Brewster kept "The White House," another popular rest stop along the Edmonton–Calgary Trail. The land is now farmed by James Brewster Jr. Near Innisfail, the railway crossed the Edmonton–Calgary Trail just north of a house known as "Constant's" or "Constantine's."

Four-and-a-half miles north of Innisfail and two miles west of present-day Highway 2A, Jack and Edward Miller ran a sod-roofed shack called "The Spruces," otherwise known as the "Half-Way House." Over the years, as the site changed ownership, the house decayed but was always restored. The building was eventually moved to the Innisfail Agricultural Grounds where it is operated as a small museum by the Innisfail Historical Society.

Red Deer Crossing

A cairn, unveiled in 1951, commemorates the old Red Deer Crossing, three and a half miles upstream from Red Deer on the site of the first trading post on the Edmonton–Calgary Trail. Near the south side of the Red Deer Crossing, G.C. King's store was the first permanent place of business established in central Alberta. Today the cairn rests on the site of the old store.

A replica of the earlier Fort Normandeau, built in 1885 by the Winnipeg Light Brigade during the Northwest Rebellion, stands one mile south of the Highway 2 Red Deer River Bridge, about

ROYAL MAIL AND STAGE LINE making fortnightly trips between Cal gary and Edmonto Stage leaves Calgar on Thursday morning June 12th and every alternate Thursday following until further notice. Makes close connection with the C.P.R. train leaving Winnipeg on the previ ous Monday morning. Leaves Edmonton Thursday morning June 19th and each alternate Thursday following. All express matter addressed in care of the undersigned will be forwarded without delay, and the advance charges paid by us. Rates, 10 cts. a pound from Calgary. LEESON & SCOTT, mail contractors, Calgary.

Advertisement for the stagecoach service on The Calgary–Edmonton Trail. From The Edmonton Bulletin*, 1885.*

one-half mile west of the highway towards the provincial campground. Some of the original timbers are a part of the rebuilt structure. One is inscribed with the names of two soldiers of the 65th Mount Royal Rifles. The fort was originally located about 100 yards south of where a cairn stands near the Red Deer Crossing. The trail ran nearby.

Red Deer to Wolf Creek

From 1886 to 1891, Sage Bannerman operated a ferry across the Red Deer Crossing. In 1889, when he brought his family west, he built a permanent home on the crossing's north side.

Along the north side of the Red Deer Crossing to Wolf Creek, opposite Fort Normandeau, the Red Deer Indian Industrial School was operated by the Methodist Church. The school housed about 150 children, and was composed of a large brick building, a barn, a workshop and a number of small houses. The school was closed during World War I and left to ruin. South of where the school stood, on a knoll overlooking the Red Deer River, there was a cemetery for the school children who died during thc scarlet fever epidemic that scourged the prairies. The mounds and crosses are now obliterated.

Wolf Creek to Peace Hills

An historic sign located along Highway 2 one mile south of Ponoka pays tribute to Fort Ostell, constructed during the Northwest Rebellion of 1885. A war party of Crees had attacked a Hudson's Bay Company post at the Battle River Crossing. In response, a battalion of the 65th Mount Royal Rifles led by Capt. Benjamin Ostell occupied the store and fortified it, strengthening its walls and digging a moat around its circumference. Though the troops at Fort Ostell never engaged in actual combat, the fort offered local residents peace of mind during the uprising.

Peace Hills to Edmonton

Six miles north of Wetaskiwin, Sam Lucas operated the first experimental industrial farm in the Northwest Territories. In 1883, the farm became headquarters for a police detachment. The farm was to be moved to Hobbema where Lucas' son was to reside as an Indian Agent, but this was delayed by the Riel Rebellion. Captain Ethier and the 65th Mount Royal Rifles constructed Fort Ethier from a timber blockhouse, along which the Bigstone Creek flowed on one side and the Calgary–Edmonton Trail across the front.

The fort, however, did not see action. Its lands were homesteaded by Lucas who ran a stopping house. The stage from Edmonton regularly made a first night's stop at the Lucas house. In 1935, a concrete foundation was poured for the blockhouse which became a museum on the Lucas homestead about five miles north of Wetaskiwin.

Passing the Leduc telegraph station, the trail terminated short of the North Saskatchewan River Hills, just beyond and to the east of the forks on the old trail that followed the south bank of the river to Fort Saskatchewan.

Of course, the Calgary–Edmonton railway meant the decline of this noble trail until its resurrection with the advent of the automobile. During its dormancy, it was used by occasional freighters and new settlers who couldn't afford railway tickets. In its new incarnation, Highway 2 continues to be known as "the trail".

Athabasca Landing Trail

Deeply steeped in Indian tradition, Athabasca trails have traditionally linked the Saskatchewan River basin with the rich lands where the Peace, Athabasca and Mackenzie rivers flow. Fur traders coming up the Saskatchewan from York Factory on Hudson Bay adopted these trails as links to Lesser Slave Lake, Fort Chipewyan, Fort Vermilion, St. Mary's near the Peace River and Fort Dunvegan.

For over 100 years, the traders' primary route to the Athabasca and Peace River basins took them up the Churchill River (either directly from Hudson Bay or by portaging from the Saskatchewan at Cumberland House) to Lac La Loche. From there it was a 13-mile slog across the marshy Methye Portage, then down the Clearwater River to where it joined the Athabasca at present-day Fort McMurray. Another route left the Churchill River at Lac Ile-à-la-Crosse (in Saskatchewan) to forge its way up the winding and sandy Beaver River to Beaver Lake, from where it was a short portage across to Lac La Biche (see The Lac La Biche Trail). The Methye Portage route with its northerly direction arrived at the Athabasca downstream from the Grand Rapids and other tortuous portages, while the Beaver River route (although it accessed the upper reaches of the Athabasca) was shallow and unsuitable for the large northern canoes. The solution was to create an overland trail. (For a third futile route, see The Fort Assiniboine Trail.)

The Athabasca Landing Trail was a portage between the North Saskatchewan River at Fort Edmonton and the elbow of the Athabasca River, or the Athabasca Landing as it was then known. Cut to allow cart passage as early as 1877, the trail was continually upgraded. The terrain between Edmonton and Athabasca Landing was fairly dry with little forest to cut. The Landing Trail must have seemed like a godsend, just one of many from the all-powerful Hudson's Bay Company. Given that the Canadian Pacific Railway would within a year reach Calgary, and that Edmonton — already a crucial trading settlement linking the north with the prairies — would become a center of commerce, the Hudson's Bay Company improved the trail in 1882. For the next several decades, all northbound commodities and people passed over this trail, including government employees, surveyors, traders, missionaries, North West Mounted Police, homesteaders and Klondike-bound prospectors.

During the Klondike Gold Rush, the trail became Alberta's first dominion highway and for 40 years it was the main transportation corridor to the far north. Fort Edmonton was at the crossroads of all trade west and north. Before the Athabasca Landing Trail was blazed as a Hudson's Bay Company trunk route, its general orientation was already in use. The Hudson's Bay Company had a post at Athabasca Landing as early as 1848. Later, northbound freight passed along the south Carlton Trail from Winnipeg, crossing the North Saskatchewan River at Joseph Lamoureux's ferry and then striking north following what is today Highway 37.

Joseph Lamoureaux was impressed by the profits to be made in serving this rough trail and had opened the first private ferry service across the North Saskatchewan near his farm at the mouth of the Sturgeon River. After 1887, with the arrival of the North-West Mounted Police at Fort Saskatchewan and the development of the trail to Athabasca Landing by the Hudson's Bay Company, business boomed as settlers made their way north. The Lamoureux family (by then the whole family had moved west from Quebec) operated a steamboat service as well. Lamoureux also established several of the first grist-mills near the Edmonton area, including

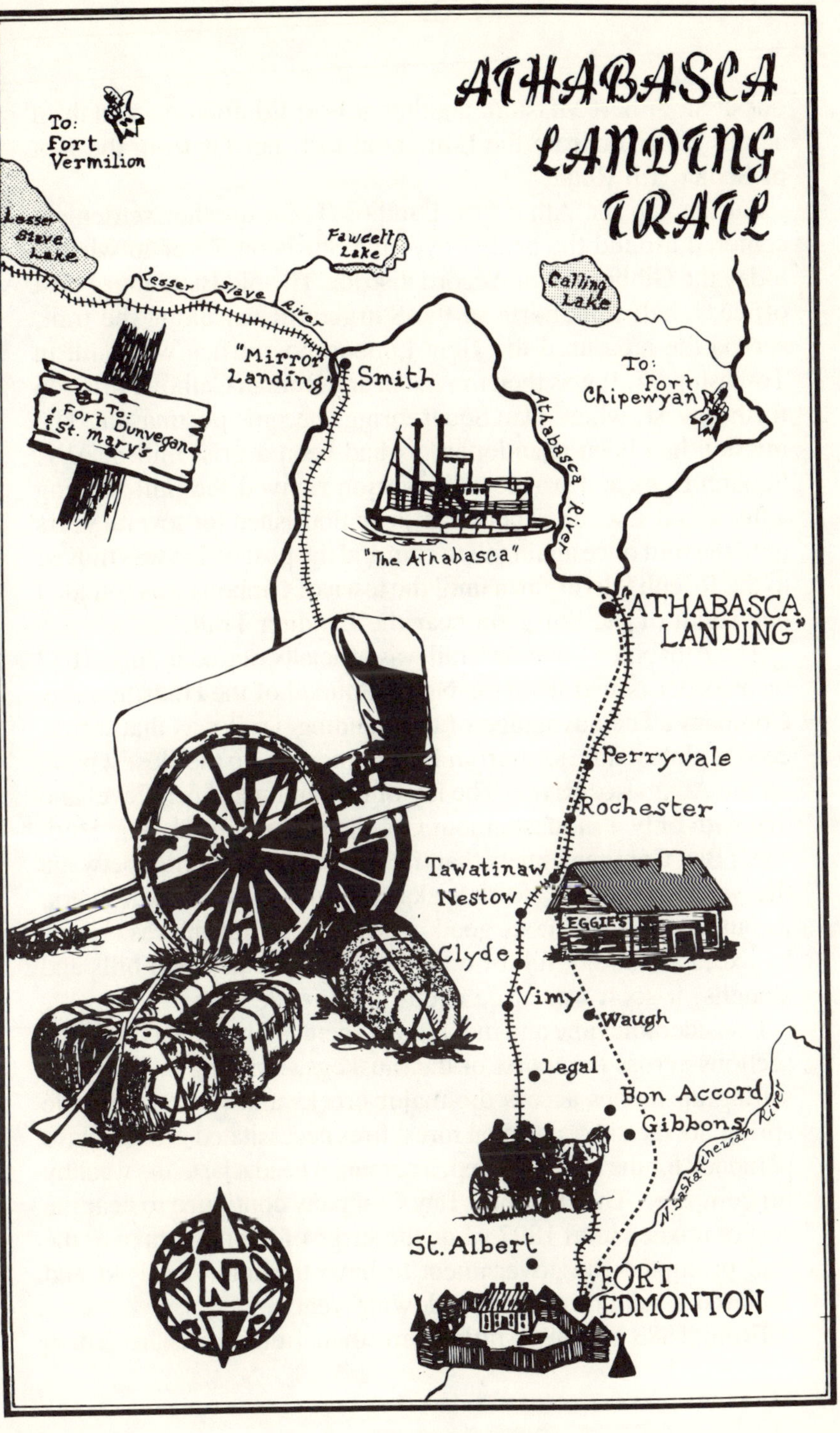
ATHABASCA LANDING TRAIL
To: Fort Vermilion
Lesser Slave Lake
Fawcett Lake
Lesser Slave River
Calling Lake
"Mirror Landing" Smith
To: Fort Dunvegan & St. Mary's
To: Fort Chipewyan
Athabasca River
"The Athabasca"
"ATHABASCA LANDING"
Perryvale
Rochester
Tawatinaw
Nestow
EGGIE'S
Clyde
Vimy
Waugh
Legal
Bon Accord
Gibbons
N. Saskatchewan River
St. Albert
FORT EDMONTON
N

one at St. Albert Mission, another at Fort Edmonton and a third across the river from the Edmonton fort, not far from the foot of Scona Hill Road.

Further up the Athabasca Landing Trail, another settlement centered around the bridge over the Sturgeon River in what is today the Gibbons-Bon Accord district. The old Battenburg post office, a half-mile north of the Sturgeon River along the trail, served the area until the New Lunnon post office was built in Township 55. It was then moved near a Roman Catholic mission further west, where Dan Squarebriggs became postmaster. The mission had been abandoned, as had been a grist mill nearby. Squarebriggs and partner Kit Watson revived the mill, adding a flume and sawmill. Their enterprise flourished for several years until the mill once again fell to ruin and the post office was moved to the Joseph Oliver farm until the town of Gibbons was founded just south of the Sturgeon near the Landing Trail.

The Athabasca Landing Trail was officially blazed in July, 1877 by an order issued from the North Council of the Hudson's Bay Company. The advantage of the Landing Trail was that it proceeded almost due north from Fort Edmonton, to the closest point on the Athabasca River. The lay of the land was fairly level and dry with only a modest amount of clearing required. The Hudson's Bay Company knew that if it blazed a smooth trail between the waterways of the North Saskatchewan and Athabasca rivers, it could more easily move goods in less time. The trail was cleared by axe, and though full of potholes, tree stumps, sand hills and sloughs, it soon wore to a deep-rutted track.

Considerable time and money was spent laying down corduroy sections across the worst of the muskegs, building bridges and setting up ferries across the major creeks and rivers. Periodic spring floods and occasional forest fires necessitated great outlays of money for the trail's upkeep, a perennial headache to the wealthy fur company. The Hudson's Bay Company continued to bear the cost of upkeep until 1897 when the surge of Klondikers over the trail prompted the government to have the trail surveyed and declared a public road the following year.

From 1883 to 1891, little more than freight was hauled to

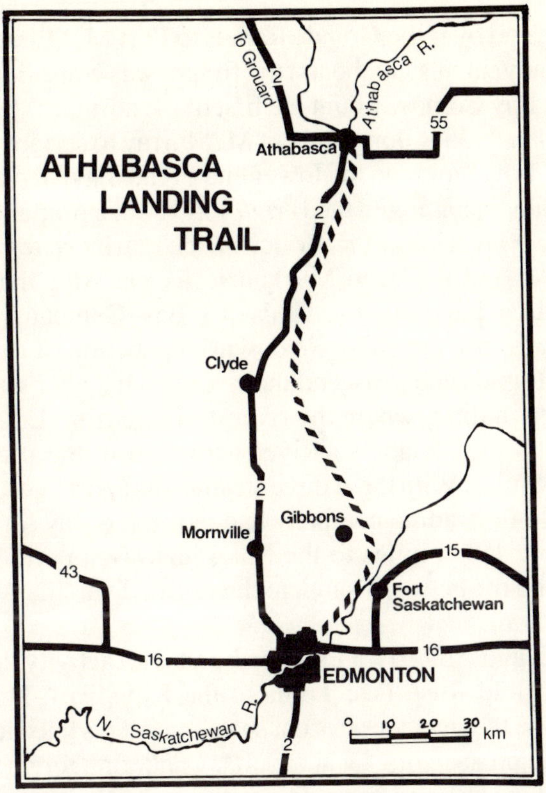

Edmonton from Calgary and up the Athabasca Landing Trail; but in 1891, when the Calgary and Edmonton Railway joined the two cities, homesteaders came in droves to Edmonton, the new and thriving gateway to the north. Athabasca Landing was the distribution point for the Lake Athabasca region and also for the Peace River district. Freight was loaded on scows that were pulled, poled, and pushed upstream to Lesser Slave Lake, and across the lake to where Grouard is today (a dozen or so miles north of Highway 2). From there freight was portaged nearly a hundred miles to the Peace River Crossing (now Peace River), then shipped downstream to Fort Vermilion or pulled upstream to Fort St. John and Fort Hope in the Peace River Country.

To reach the hinterlands of the Lake Athabasca region and further north to the Mackenzie River, goods were floated downriver on scows, and over the Grande Rapids to Fort McMurray, where (with the advent of steamboats) the freight was consolidated. The Hudson's Bay Company built the first of the northern steamboats, the *Grahame* (135 ft. long) at Fort McMurray to service the Slave River. At Fort Smith on the lower end of the Slave River rapids, the company inaugurated the *Wrigley* in 1887, a propeller-driven boat some 90 feet long, for service on the murky waters of Great Slave Lake and the deep Mackenzie River. Also that year at at Athabasca Landing, the Hudson's Bay Company built the *Athabasca* to ply the 167-mile downriver section of calm water to Grand Rapids and to ascend the river to what later was known as Mirror Landing, where the river draining from Lesser Slave Lake meets the Athabasca River across from the present-day hamlet of Smith. With these three steamboats and a fleet of scows, the age-old fur trading company had paved the way for summer travel some 1,800 miles to the Mackenzie Delta on the Arctic Ocean. And the network's link to the railroad was the Athabasca Landing Trail.

The trail and Athabasca Landing burst with activity during the Klondike Gold Rush (see Trails to the Klondike). During the winter while the river was frozen, shacks and tents lined its shores. Miners and prospectors, local experts, wandering fortune-seekers and even businessmen were anxious to stake their claims while the getting was good. Once the snow thawed, the tents disappeared with hardly a trace. Anything that could be made to float was put into the river.

Then between 1906 and 1912, thousands of settlers headed to the Peace River Country along the Athabasca Landing Trail. From 1909 to 1914, as many as 100 boats and barges were built and launched each season, and no less than four sternwheelers carried the freight and new settlers. But as the railway penetrated the northwest, the steamers retreated, or decayed alongside the waters they had once navigated. Despite the feeble attempt of the Edson–Grande Prairie Trail to usurp traffic to the Peace (see The Edson–Grande Prairie Trail), the pace of movement — either by wagon or rail — swelled over the Athabasca Landing Trail. By September of 1912 the Canadian National Railway was whisking settlers to

At Athabasca Landing, traders and others load their goods onto scows for distribution up and down the Athabasca River.

Athabasca Landing three times a week. Finally, however, Athabasca Landing lost its status as Gateway to the North when the Edmonton, Dunvegan and British Columbia Railway reached the more navigable Peace River.

There were actually two Athabasca Landing trails: one for winter and the other for summer travel. (They were also known as the high and the low road.) At first, the 100-mile route was only a winter road, but by the early 1900s, summer cart traffic routinely used the Hudson's Bay Company's bridges and ferry scows.

One trail to Athabasca Landing exited Edmonton by way of the St. Albert Trail past Morinville, Peavey, Legal and Fairydell; the second departed by way of the Old Victoria Trail (Fort Road) following its northeast route to the Sturgeon River (described above), to about one and a half miles from Gibbons.

The winter trail from Halfway Lake followed the lakes and creeks to another stopping house at Stony Creek (from an Indian expression meaning "a river between the hills"). From Stony Creek the trail meandered through hills passing through Rochester and skirting the campsites at Two Creeks before it dipped into the valley to G. Luve's stopping house (now Perryvale). From there it followed the creeks and traversed the rivers to Athabasca Landing. (The name "Athabasca" comes from a Cree word meaning "Where there are reeds," describing the river's marshy delta. The Cree name for the townsite meant "landing").

The summer trail lay from Edmonton to Fort Saskatchewan, then veered north between Oliver and where Highway 37B intersects the Fort Trail. It followed the Sturgeon River through the town of Gibbons. A water tank in Gibbons stands on the original trail. From Gibbons, it crossed the ford at the present-day railroad trestle, where once an eatery and inn served travellers.

In the spring of 1898, as the Klondike rush reached its peak, J. M. Kennedy began operating the first stagecoach of the Northern Transportation Company service over the trail, whisking patrons to and from the Athabasca Landing and between the many stopping houses situated along the way. It seemed as though every homestead adjacent to the trail offered some type of accommodation (though some were more accommodating than others).

Egge's stopping place, one and one-half miles south of the Bridge Lakes bridge between Waugh and Clyde, was a sight for sore eyes on cold nights. By all accounts, Mrs. Egge — no doubt a rosy-cheeked motherly type — served up generous portions of eats and fun. Edna Shore, an Edmonton author, wrote the following in "Athabasca Landing Trail," a delightful account of her stop at Egge's in 1910 written for the *The Alberta Historical Review:*

> When I went inside, the kitchen with its long table of goods, had a delicious smokey smell and the cakes and pies seemed as big as cartwheels. . . . In a little while the air became warmer and, with the sounds of laughter, more cheerful. . . . The piano opened its mouth to show its teeth in a pleasant smile and to join in the singing of "Little Red Wing."

DAVID CUNNINGHAM

Today, only a cairn marks the site of the original Athabasca Landing in the town of Athabasca.

A replica of Egge's has been constructed at Fort Edmonton Park in Edmonton.

Another stopping place, that of John Milligan (located one and one-half miles east of Bon Accord), afforded an equally grand and affable arrangement. In 1901, the Milligan's built a storey-and-a-half log house — 20 by 24 feet — with four bedrooms, a living room, a 12-by-24 foot kitchen (prairie folk lived in their kitchens) and a pantry. Later, an immense bunkhouse with a slab roof was added to accommodate the many freighters passing by. As many as 40 men would sleep on the floor and in the haylofts, their horses tethered within a nibble's reach. These men were a hearty sort who worked when they worked, but played when they played. In the evenings, after their suppers of fried dough, reheated beans, condensed milk and occasional sardines, they would gather around the campfire, sing songs and spin yarns that would make a fisherman blush. Sometimes the mounted police would happen along to add their two cents' worth of fun and handcuff the children for being naughty. One policeman even brought along the evidence of a murder in a gunnysack — a severed head!

At Halfway Lake (aptly named as it was halfway to Athabasca Landing along the trail) Mr. Pat Nugent owned and operated another stopping place, Halfway House. Bud Egge, the keeper of Eggie's, described a typical evening scene at the house:

> A cold winter night and the barns and outbuildings filled to overflowing with oxen and horses. The house heavy with tobacco smoke as a motley gathering of police, trappers, traders and government officials sought refuge from the cold.

Between 1901 and 1905, the lonely trail became fairly crowded with new places where freighters and their teams could rest and feed, sheltered from the elements.

The first settlement along the Athabasca Landing Trail was established in 1887 when Johnny Gullion opened his place three miles south of the bridge between Bridge Lakes. Ten years later, surveyor J. K. McLean reported the presence of only four new sites north of the Sturgeon River.

Ed Williams ran a house one-and-a-half miles south of Athabasca Landing. There was a house situated on the south side of Stony Creek one mile east of today's Tawatinaw. The Wabiscas owned a house and stables at Pot Lake, three miles north of the Bridge Lakes bridge. A man named Bluff erected his house on the south side of the Redwater River crossing in 1893.

Newton (Bud) Egge settled his place, perhaps the most famous of all the houses, in 1898, one-and-a-half miles south of the Bridge Lakes Bridge.

Some of the local residents east of Bon Accord built bridges across the Sturgeon River in hopes of diverting traffic past their homesteads. Tom Lewis ran his house, Three-quarter House, east of today's Perryvale. John Milligan, already mentioned, operated near where a short cut over the Sturgeon River south of his place joined the original trail. Joe Rye's place, not on the original trail, was situated a mile and a half east of where the road allowance was already being used.

Other stopping houses included (from north to south):

Billy Smith's on the west bank of the Tawatinaw River, a mile north of present-day Meanook;

Wabisca's on the Creek Road just a mile south of Meanook;

Frank Robert's at Sandy (Spring) Creek half a mile east of Perryvale;

Hunter's stopping house at Two Creeks, two miles southeast of today's Rochester;

Ed William's place on the south side of the Redwater Bridge near the house owned by Bluff;

Widow Leclair's (also known as Bob Shaw's) at the north end of Lily Lake.

Preservation of the Athabasca Landing Trail

The Athabasca Landing Trail is alive and well today. Long segments are preserved in a natural or near-natural condition through private land and parallel to public access roads. Historic sites along the trail include former stopping houses, campsites, bridges and ferry crossings, and trail junctions — landmarks which could be suitably identified with signs and markers.

In 1975, the Trail North Foundation undertook a major project to mark this important road. Assisted by student labourers from towns along the trail, they cleared bush, erected signs and stiles and planned parks and interpretive facilities to augment the experience of travelling the trail.

Much of the trail at its south end from Edmonton to Bon Accord, passes over marhlands or is obscured by vegetation (which helps to preserve wagon ruts), but about half of the remaining portion from Gibbons to Athabasca has been cleared and marked. In this way, one of Alberta's most important post-railroad historic trails is still accessible in a preserved, unimproved state.

Rocky Mountain House

The North West Company surely regretted building the first Rocky Mountain House, for it was surrounded by muskeg and warring Indian tribes, food was scarce, famine frequent, and the failing enterprise was expensive to keep.

But the fort had its purpose. Not only was it built as a trading post to deal with the Indians, but also as a springboard to the unknown lands across the Continental Divide and south into the prairies. In their competition for new markets, the trading companies had pushed westward until by the mid-1780s, they had reached the vicinity of what is now Edmonton. By then the Nor' Westers, having dangerously over-extended their supply line to Montreal, needed a base if David Thompson was to continue his mission of exploring the mountains for a western passage. Rocky Mountain House was that base. Nearby, the Hudson's Bay Company, neck-to-neck in the fierce race west, opened Acton House for similar purposes.

Rocky Mountain House was constructed on the upper reaches of the North Saskatchewan River about 50 miles west of Red Deer and three miles from its present-day namesake. Though called Rocky Mountain House, its proximity to the Rocky Mountains was a distant 50 miles.

The early years were turbulent for the Rocky Mountain House traders, whose victories were rare, failings frequent. Facing death threats, eating dog meat when times were lean, and waiting for their Blackfoot clients became the daily ritual for these hearty men.

Wrote Thomas Woolsey, in a letter to Wesleyan Missionary Notices dated November 9, 1858:

> Through the great scarcity of provisions, we have all been on short allowance for some days past. Some French Canadians, whose ravenous appetites could not be restrained, have devoured two dogs. This has arisen out of Indians not bringing dried meat to the Fort as soon as expected. I am informed that some of Company's posts there depend almost entirely upon the Farm fishery, which involves starvation if it proves a failure. At one of these posts is a small island, just opposite the establishment, where there are graves of men who died of starvation. Religious services have been conducted every evening during our stay.

Over its 75 year history, Rocky Mountain House was burned twice and abandoned a number of times.

It is the Blackfoot who deserve the credit for the post's founding. In 1799, a band of Peigans (part of the Blackfoot Confederacy) intercepted a party of traders comprised of men from the North West and Hudson's Bay companies. These Indians guided the traders along existing trails to the founding site. The Peigans immediately liked the locale, for it was close to their hunting grounds and far from the territory of their sworn enemies the Assiniboines. So with approving nods, the Hudson's Bay Company traders built and the eager Blackfoot vowed their patronage. The Hudson's Bay Company traders, determined to thwart their competitors' efforts, opened nearby Acton House.

But if the Blackfoot seemed then engaged by the prospect of trade, they failed to act. Doomed from the start, the post was closed in 1802 because the Blackfoot were more concerned with policing their territory than trading furs, and the Kootenay Indians of British Columbia who might have traded, were loathe to come over the mountains lest they be dealt with severely by the hostile Blackfoot.

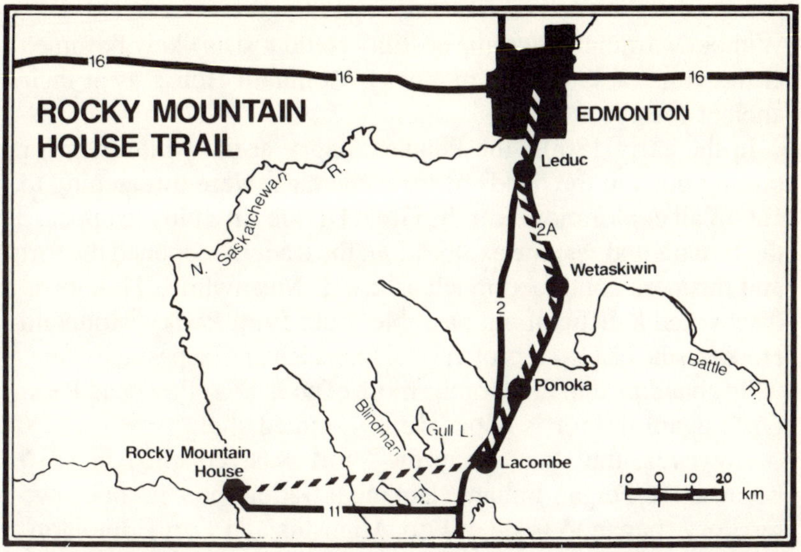
ROCKY MOUNTAIN HOUSE TRAIL
EDMONTON
16
Leduc
2A
2
Wetaskiwin
Ponoka
Lacombe
Rocky Mountain House
11
N. Saskatchewan R.
Battle R.
Blindman R.
Gull L.
10 0 10 20 km

THE ROCKY MOUNTAIN HOUSE TRAIL
FORT EDMONTON
N. Sask. R.
LEDUC
Bow
Millet
Wetaskiwin
River
Hobbema
Battle R.
Ponoka
Ft. Ostell
Morningside
Trail
Lacombe
Wolf
Blindman
Gull L.
Medicine
N. Saskatchewan R.
ROCKY MOUNTAIN HOUSE
Wolf's Trail
Sylvan L.
R.
Red Deer R.
D.MARKO 85

Without warning, that same hostility could just as likely be aimed at the vulnerable traders of Rocky Mountain House as at their ancient foes.

In the early 1800s, the Peigans, angry at the white men for putting guns in the hands of their enemies, were threatening to cut off all exploration over the Great Divide. In a ploy to appease the Indians and resume exploration, the traders reopened the fort and business continued much as usual. Meanwhile, Thompson discovered a difficult but passable route from Rocky Mountain House to the headwaters of the Columbia River — a passage eventually abandoned in favor of the more expedient Yellowhead Pass. Once again the fort's doors were slammed shut.

However, the "never say die" fort was reopened in 1810 because the Peigan Indians, lacking a trading post in their own territory, began to trade at Fort Augustus and Fort Edmonton. As the Crees also traded at these forts, they and the Peigans, allies in the Blackfoot Confederacy, were at imminent odds and poised for war. The North West Company and the Hudson's Bay Company wisely decided to reopen their houses, thus providing the Peigan Indians with a place to trade — the Crees out of the picture.

With the amalgamation of the two companies in 1821, Acton House was closed and Rocky Mountain House was chosen as the more substantial base of operation. Throughout its remaining history, Rocky Mountain House was closed and moved a number of times, but was always reopened during the winter months at its original site because the persistent Blackfoot continued to interfere in one way or another. Either all hell broke loose when the traders attempted to barter with Blackfoot enemies, or the American whiskey merchants were stealing business from the Canadian traders. In 1875, the tired old fort was abandoned for the last time when the Hudson's Bay Company sold its interests to the Canadian Government and the Indians were moved to reservations.

With time, little remained of Rocky Mountain House but two of its chimneys which have since been reconstructed. A park near Rocky Mountain House displays replicas of the original buildings and pioneer memorabilia.

Read Hugh A. Dempsey's book, *A History of Rocky Mountain House*, for a thorough historical account of Rocky Mountain House. Also in the book, William C. Noble details the excavation and historical identification of the fort's structure and artifacts.

The Trail to Rocky Mountain House

The Rocky Mountain House Trail is likely centuries old, long used by the Peigan and Blackfoot in their nomadic hunts. Certainly the latter-day natives used the trail, for it was over that route along what is known as Wolf's Track, that the Blackfoot had led the founding traders.

There were three routes to Rocky Mountain House. In its early years, supplies were shipped on the North Saskatchewan River. The overland trail, later used, was composed of two segments: the first track, known as the Bow River Trail, more or less followed the Edmonton–Calgary Trail as far south as the Gull Lake region. Passing between Whitemud and Blackmud creeks, it skirted the east side of Wetaskiwin (Peace Hills) and Hobbema (Bear Hills) and then crossed the Battle River near Ponoka, continuing on close to Lacombe. At this point, the trail split and the second segment veered west, crossing the Battle, Medicine and Wolf (now Blindman) rivers via the south shore of Gull Lake and Horse-Pound Creek, a tributary of the Medicine River. This was known as Wolf Track or Wolf's Trail. It experienced principally winter use by dogsled, and was deeply mired with bogs and wagons ruts.

Doctor Hector of the Palliser expedition referred in his journal to having travelled this segment of the Rocky Mountain House Trail. Moreover, Peter Pruden of Buckingham House noted on June 17, 1799, that "three more Canadians with some Indians went to a place the Indians call the Wolf's Track, which is along a narrow plain that leads down to the river opposite to where they are going to build." He was no doubt referring to the founders of Rocky Mountain House.

Over the years, some notable figures trekked the trail to Rocky

Mountain House, including Anthony Henday. The second recorded passage was that of Peter Pangman, one of the founders of Rocky Mountain House. David Thompson later journeyed from Fort Augustus to Rocky Mountain House by way of the trail, crossing the Battle River just east of Bear Hills (near Hobbema) and following Wolf Trail across Wolf Creek to the Clearwater River about two miles above its mouth. (A sign along Highway 11 just east of Rocky Mountain House honours David Thompson, fur trader, surveyor, geographer and one of Canada's greatest explorers.) In 1810, explorer Alexander Henry the Younger found the first part of the trail to be well travelled and in his words, "as broad as a wagon road." However, he thought that the second segment, especially after the trail crossed the Blindman River (Wolf River), would have been especially "horrid" during rainy seasons.

Henry's interesting journal, dated October 4, 1810 (when he was in charge of Rocky Mountain House) gives a vivid account of life on the trail to Rocky Mountain House:

> At eleven we set off, still up the small (Horse Pound) river until we crossed it, when we immediately entered the strong woods, and continued through an ugly, gloomy country covered with pine and aspen, poplar, willow, and long grass, with a crooked path to avoid the several masquegues (muskegs).
>
> After passing a thick wood for several miles we came out into the small plain where lies the Big Stone, and soon fell upon the old Peigan road which leads to the establishment. Our course then turned S.W. and W., through small prairies and pieces of thick woods, mostly of the Rocky Mountain pine. We soon crossed Riviere a l'Eau Claire, which is here 80 yards wide and has a strong current. The water is so clear that I was deceived in its depth, and narrowly escaped getting wet. Here we perceived some fresh horse tracks, coming and going which led us to suppose that the natives had been at the old house to see if we had arrived; for we gave them to understand last spring that peole should winter here for the Peigans. The track led

> down Riviere a l'Eau Claire, among small prairies and pines, to the Saskatchewan. Upon coming in sight of the old fort, we were surprised to see some people standing at the gate, and smoke arising from within the stockades; on nearer approach our surprise increased when we observed them to be whites and Indians. The curent was too swift to cross opposite the house, where there is a strong rapid; we therefore passed about 1/4 of a mile above, to the head of the rapids where we crossed with ease. On coming down to the fort we found the Columbia brigade had been in possession since the 24th ult.

After Rocky Mountain House was closed in 1875, traffic on the trail segment west to the Bow River was greatly reduced. It wasn't until the influx of settlers penetrating the hinterland that the trail to Rocky Mountain House (or House Trail as it was also known) came alive once more. By 1908, the vicinity of Rocky Mountain House could be reached by roads and trails from Lacombe, Red Deer and Innisfail. In that same year, the population near Rocky Mountain House consisted of about eight to ten families.

The road leading west from Red Deer had long been used by traders and Indians. It lay in a rough hewn, winding fashion on the highest land to Sylvan Lake, which was crossed when the surface was solidly frozen, but was otherwise avoided. The trail then veered across country to Eckville from where it passed through the Blueberry Hills to Leslieville. There, for many years, Dan McLeod ran a stopping house.

A few portions of the Rocky Mountain House Trail are preserved. A section parallels the Canadian Pacific Railway tracks south of Highway 12, just east of Gull Lake. Traces also remain at the Lobstick River crossing near Leslieville and at the Blindman River crossing just south of Bentley. The latter section — approximately a quarter of a mile long — lies on the property of Louis Cabelka. Both ends of the trail are made accessable by a secondary road which renders public use of the trail possible without trespassing on private property. This segment of the trail — which lies on a land ridge unsuitable for cultivation as its soil is gravelly and used for pasture — has excellent potential for full recreational and interpretive development.

Lac Ste. Anne Trail

For almost a century the trail from Edmonton to Lac Ste. Anne actively served to unite the surrounding communities. Lac Ste. Anne had long been an important fishery for the Hudson's Bay Company and source of provisions to fur traders who might otherwise have suffered meager diets during the long winters. It grew in the 1800s to become a major Metis colony and was the nexus for many criss-crossing paths.

Until 1900, the Hudson's Bay Company used a boat at the lake's narrows to cross from the south to the north side of the lake. Of course the horses had to swim across, so they were taken to the west part of the lake, opposite from Horse Island. The horse trail emerged from the lake's north shore and connected with the trail leading from the narrows.

The Lac Ste. Anne Trail, a branch of the Fort Assiniboine Trail, came into being in 1844 when cut by Father Thibault to his new mission at Lac Ste. Anne. Father Thibault's Roman Catholic mission was one of the earliest in Alberta. Thibault, an oblate missionary from St. Boniface at Red River, hoped to introduce the Metis to a peaceful life of farming divorced from the warring of the plains and the nomadic toil of the hunt.

A member of the Palliser expedition visited the priests at Lac Ste. Anne in the winter of 1858 and found two settlements of about 80 people each making their living from cultivating the land and fishing in the lake. Father Lacombe, however, later found the land around Lac Ste. Anne to be unsuitable for large farming projects so he moved the mission to St. Albert. A cairn marks the Thibault mission site at Lac Ste. Anne.

Father Lacombe's biographer, Paul Emile, noted that the Lac Ste. Anne Trail was one factor in Alberta's development of organized travel by Red River cart. Over the years, independent traders built a number of posts in the area, so that trails criss-crossed from every direction. Many of these trails were marked only by blazings on solitary trees.

Though by 1885 the Canadian Pacific Railway had made many trails in the area obsolete, the Lac Ste. Anne Trail saw continued use because it afforded access to the vast area between the Athabasca and Peace rivers, and to Fort St. John and points beyond, where men roamed in quest of undiscovered fortune. The trail also accommodated the many survey crews and scientific expeditions being sent by the government to map and quantify the potential of the west. Local pioneers and Metis residents who had spent their lives roaming the nearby woods, trapping furs or panning for gold, guided these parties — and later the Klondikers — over the confusing network of trails.

As homesteaders drifted into the surrounding country, Lac Ste. Anne evolved as a hub through which various trails branched. There was, for example, a vague wagon trail from the Lac Ste. Anne settlement to the lake's narrows where an old Stoney Indian by the name of Peter Painted Stone (nicknamed Wee Zaw — French slang for "Bird") ran a ferry of sorts. The ferry, made of barrels and planks, could float about a ton per trip, and was moved by state-of-the-art woman power. Wee Zaw lived in a lodge at the edge of Lac Ste. Anne, and though he took the credit for being the "brains" of the operation, it was really his wife and several other women who laboured like mules. He would stand dry on the opposite bank and boss them around like a little French general.

In 1907, the government put Wee Zaw's ferry out of commission and replaced it with a more efficient model built by Edmonton-area pioneer John Walters. Also in that same year, the government cut a trail from the Narrows westward across the Alexis

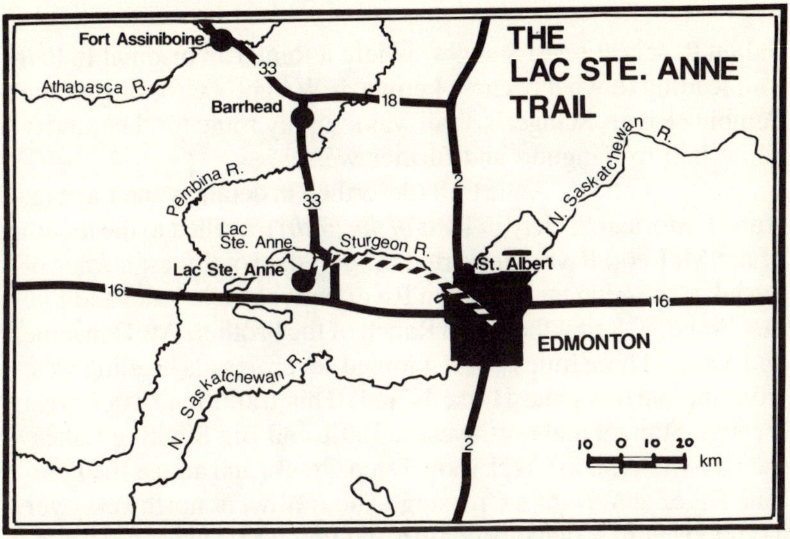
THE LAC STE. ANNE TRAIL
Fort Assiniboine
Athabasca R.
33
Barrhead
18
2
N. Saskatchewan R.
Pembina R.
Lac Ste. Anne
Sturgeon R.
St. Albert
Lac Ste. Anne
16
EDMONTON
N. Saskatchewan R.
10 0 10 20 km

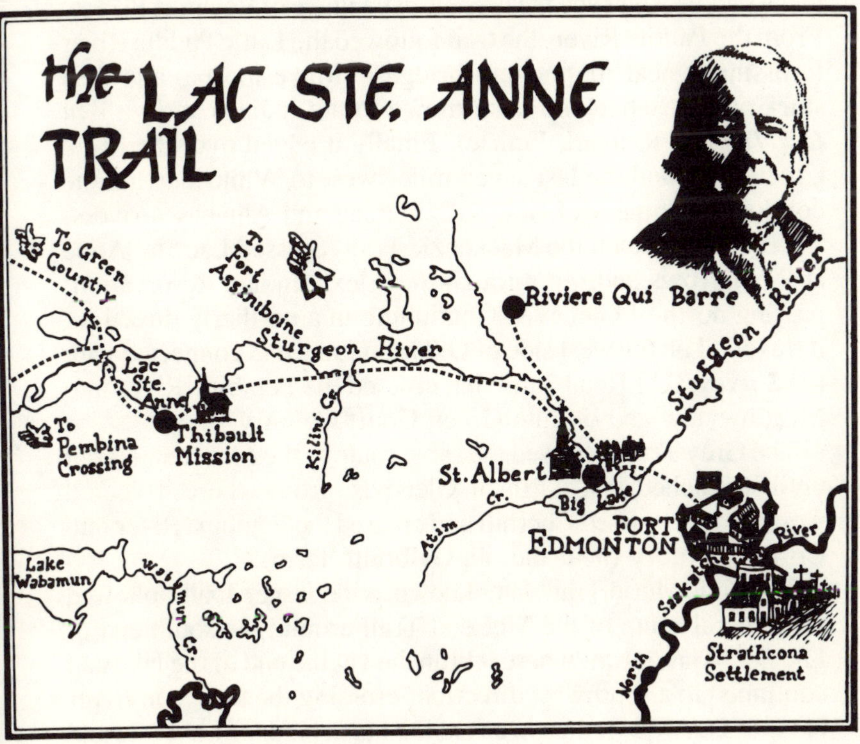
the LAC STE. ANNE TRAIL
To Green Country
To Fort Assiniboine
Sturgeon River
Riviere Qui Barre
Lac Ste. Anne
Thibault Mission
To Pembina Crossing
Kilini Cr.
St. Albert
Big Lake
Atim Cr.
Sturgeon River
FORT EDMONTON
Lake Wabamun
Wabamun Cr.
North Saskatchewan River
Strathcona Settlement

Indian Reservation to Stanger, where it joined an original Indian trail leading to what became known as Wright's Crossing on the Pembina River. Stanger's Trail was a supply route for the railway being laid to Sangudo and further west.

Another Lac Ste. Anne trail (described in detail by the Lac Ste. Anne Historical Society in *West of the Fifth*) travelled to the mouth of the McLeod River. It skirted around the southwest corner of the lake, crossing the Sturgeon River between Isle Lake and Lac Ste. Anne, just past the Hobo Ranch of the brothers McConachie and Yates. Three miles north it joined several trails heading west from the narrows and Horse Island. This trail then struck west through Stanger and northwest to Little and Big Stinking Lakes, then down Peeshoo Creek (now Deep Creek) and across the Pembina River at Wright's Crossing. The trail went northwest over a section cut by Cowper Rochfort and Percy Michaelson in 1905, to the Paddle River and the present-day village of Rochfort Bridge. From the Paddle River, the trail followed the Little Paddle River (crossing it near the modern bridge in Mayerthorpe) and then stuck across Selleck's Flats (known to the Indians as *Pog Win Ock Tick*, or Rhubarb Prairie). Finally it edged over the Mink Creek Flats and the last seven miles west to Whitecourt, at the confluence of the McLeod, Sakwatamau and Athabasca rivers.

Yet another road, the Mackenzie Trail, crossed Lac Ste. Anne at the narrows and ran through the Alexis Indian Reservation, passing north of Glenevis. Continuing in a northerly direction, it travelled on the west side of Old Man Lake and connected with the Survey Tote Road Trail that crossed the Pembina River into Meadowview and on into Green Court Country.

The Grey Trail followed the same route as the Mackenzie Trail until it reached just north of Glenevis, from where it veered westward and passed Cherhill and crossed the Pembina River into Cosmo territory (near the old Galbraith farm).

The Yellowhead Trail, later known as the Jasper Trail, followed the general route of the McLeod Trail around the south end of Lac Ste. Anne, from where it branched at the end of the lake and continued in a southwest direction, crossing the Sturgeon River close to Darwell. It then brushed the north bank of Lake Isle and the south side of Sturgeon Creek, veering westward for about five miles to cross the Pembina River where Entwistle now stands. It followed the south side of Chip Lake and struck west into the mountains.

Until 1858, goods bound for the Jasper area were sent by way of the Fort Assinboine–Athabasca River route. Afterwards, however, the Hudson's Bay Company assigned H.J. Moberly to supervise Jasper House. Desirous of an overland route to his new assignment, Moberly and an outfit of men left Edmonton with about 40 horses, following the trail from Edmonton to Lac Ste. Anne. From there they went to Island Lake and then to the Pembina River. Proceeding across the muskeg to Lac Brule, they finally reached Jasper House. Theirs was the first pack train ever taken to Jasper House.

The main trail to Lac Ste. Anne from Edmonton passed through the attractive country of the Calahoo Indian Reserve where the Indians lived in spacious log houses, and constructed barns and corrals which impeded the trail. Wrote Mr. J. Priestly, who homesteaded in the Onoway area:

> In the latter years I came to realize that there was a method behind all this, though, at the time, it seemed strange that there was no bridge over the shallow creek in front of the buildings which took the spring runoff to the south on its way to Big Lake. The Indians were resisting the establishment of a right-of-way through the Reserve. From time to time we would find one of the grassy open spaces on the trial plowed up and sown to oats, with a rail fence erected pushing the trail into the edge of the bush.

And yet, Chief Calahoo and his wife were very sociable to passers-by, whom they treated to good meals and comfortable lodging.

Further along, Dan Noyes ran a hotel at Noyes Crossing. Mr. John Morrison recalled the service as being a "harbour of refuge" run by a "likable gentleman." Noyes dispensed all the news and gossip he could gather from as far away as the Peace River Country, while Mrs. Noyes ensured all her guests were well sated.

Though such pleasant happenings characterized life along the Lac Ste. Anne Trail, it wasn't until the mid 1900s that settlers took stock in the land and populated the area. In 1903, Prime Minister Wilfred Laurier announced that the Grand Trunk Railway would span the country, taking the Yellowhead route across the mountains. The Grand Trunk reached Edmonton by 1909 and work on the old Jasper Trail up to Yellowhead commenced. In 1910, tracks were laid as far as Edson and by 1912, the line was operating between Jasper and Mt. Robson. As was the trend, settlement paced with the iron horse, and the Lac Ste. Anne Trail fell to disuse.

Trails to Lac La Biche

It can be said that Lac La Biche literally stood at the crossroads of Western Canada's history. When the Hudson's Bay Company was granted its charter by Charles II in 1670, a small group of elitist Londoners was given absolute right to all the fur-rich lands and rivers that drained into Hudson Bay. The fur monopoly controlled an empire that was larger than Europe. Though they were speculators, the fur barons were content to remain in London. Rather than risk sailing the stormy Atlantic Ocean, they sent agents (voyageurs) to Hudson Bay to collect a harvest of furs brought by Indians from as far away as the Rocky Mountains. For a century the hierarchy remained unchallenged: the vast northwest was never explored. The Europeans had nonetheless left their stamp on a primitive culture: the Blackfoot and other tribes of the southern plains did not rely on canoe transport and could not make the journey east. Yet they knew of the white man and his fabulous offerings.

Though the voyageurs and their descendents had not explored the lands beyond Hudson's Bay, they learned from the Indians of the great rivers and highways leading to the teeming northwestern hunting grounds. The white man's influence was setting the stage for further exploration. The Hudson's Bay Company could no longer justify its sole control of a potential nation. Moreover, the Hudson Bay charter could not legitimately lay claim to the resources from northern rivers, such as the Peace, which did not flow into Hudson Bay. Descendents of the voyageurs — courageous men such as Simon McTavish, Peter Pond and Alexander Mackenzie — organized themselves into the North West Company based at Montreal and set out to conquer a bold, new land.

In 1778, Peter Pond was the first recorded white explorer to cross the Methye Portage and paddle the Athabasca River. He built the first trading post to stand on its banks and was the first white man to build a residence and reside in what is now Alberta. Though he explored the area with the modest hope of trading with the Chipewyan and Cree Indians — by all accounts a successful venture — Pond's dream was to discover a trade route to China and the Orient. Though Pond's dream wasn't realized in his time, his exploits led the way to further western exploration by such men as Alexander Mackenzie and Simon Fraser who ventured the same route and in whose honour two of Canada's mighty rivers are named.

Peter Fidler was another of the goodwill ambassadors sent by the fur trading companies to befriend the Indians, learn their language and customs, and collect details of Alberta's geography and resources. A skilled surveyor, Fidler, in 1792, was the first geographer to accurately map the Battle, Red Deer and Highwood rivers, as well as the Rocky Mountains. By 1799, much of Alberta — its waterways, prairies, hills and mountains — had been thoroughly explored and mapped. At the time of Fidler's arrival, forts George and Buckingham House were the most westerly bases of operation until the building of forts Augustus and Edmonton House in 1795 further up the Saskatchewan River near present-day Fort Saskatchewan. The early portage route to Lac La Biche stretched from the vicinity of Fort George (North West Company), Buckingham House (Hudson's Bay Company) and Fort Vermilion (operated by both companies), all dotting the North Saskatchewan River. The North Saskatchewan was, in turn, the main artery along which exploration and settlement spread west.

According to Nor'wester Duncan McGillivray, the area was

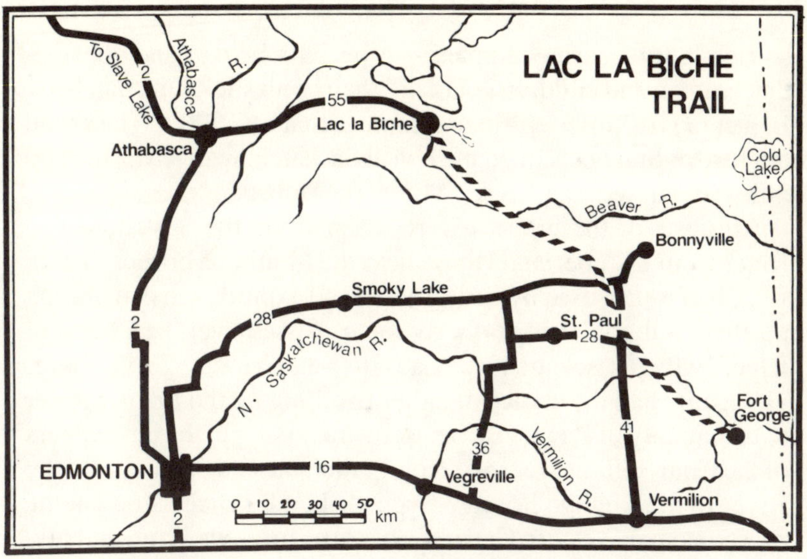
LAC LA BICHE
TRAIL
To Slave Lake
2
Athabasca
R.
55
Lac la Biche
Athabasca
Cold Lake
Beaver R.
Bonnyville
Smoky Lake
28
2
St. Paul
28
N. Saskatchewan R.
Fort George
41
36
Vermilion R.
EDMONTON
16
Vegreville
Vermilion
0 10 20 30 40 50
km
2

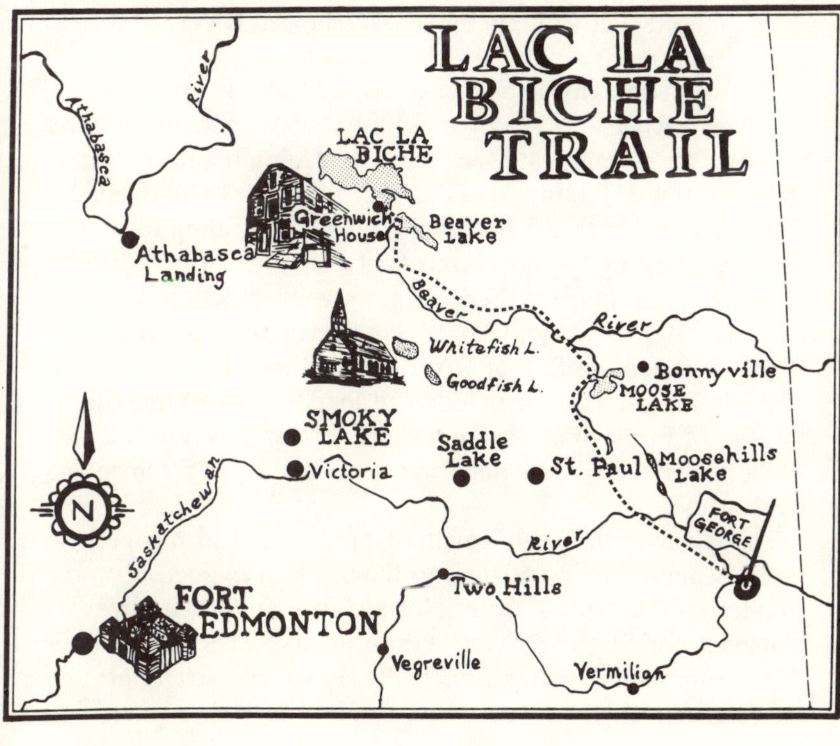
LAC LA
BICHE
TRAIL
Athabasca
River
LAC LA
BICHE
Greenwich
House
Beaver
Lake
Athabasca
Landing
Beaver
River
Whitefish L.
Goodfish L.
Bonnyville
MOOSE
LAKE
SMOKY
LAKE
Saddle
Lake
St. Paul
Moosehills
Lake
Victoria
N
Saskatchewan
FORT
GEORGE
River
Two Hills
FORT
EDMONTON
Vegreville
Vermilion

so rich in fur-bearing animals — especially beaver and otters — that women and children could kill them with sticks and hatchets. In fact, so fruitful was Alberta's larder that by 1800, no less than ten fur-trading posts lined the North Saskatchewan River for 800 miles as far west as Rocky Mountain House.

All posts of the prairies were reached via the Saskatchewan River from Cumberland House near the Manitoba border. There were two routes used by early traders and explorers in penetrating northern Alberta. The northerly route crossed over the Churchill River which ascended to Lac Ile-à-la-Crosse. From here, voyageurs had the choice of either paddling up the Beaver River to the vicinity of Lac La Biche, or by traversing to the headwaters of the Churchill, then across Peter Pond Lake and up the La Loche River to the Methye Portage. Indeed, Lac La Biche stood at the crossroads of Western Canada's history, for it was conveniently positioned on the Beaver-Athabasca and Lesser Slave Lake route to the Peace River Country and Saskatchewan River; a tentacle of exploration.

Though at times it was almost unnavigable due to lack of water, the Beaver River was heavily used for at least 25 years, serving to access the Athabasca, Mackenzie and Columbia River watersheds. While travelling to Fort Vermilion in 1814, Gabriele Franchere discovered a horsewhip on a trail leading to Moose Lake, suggesting that horse travel between the North Saskatchewan forts and Lac La Biche was common practice. In 1824, Hudson's Bay Governor George Simpson en route to the Pacific Coast, paddled the Beaver River until he found it too dry and difficult. Simpson and party followed an Indian winter trail (likely by way of Frenchman, Ironwood and Elinor lakes) to Lac Lac Biche. There he noted a rather large population of freemen and halfbreeds.

Though over the years smallpox epidemics had ravaged the prairies' halfbreed population to the brink of extinction, some small communities — fisheries and farming co-operatives — managed not only to survive, but to prosper. The residents of these communities relied principally on wheeled transport and thus blazed many of the first trails suitable for carts — including

DAVID CUNNINGHAM

The view today from the site of Fort George, the southern terminus of one of the trails to Lac La Biche.

the portage trail to Lac La Biche — around which their settlements sprang up at the crossroads.

The establishment of missions in northern Alberta influenced development of trails to Lac La Biche and other portage and cart tracks which displaced the long water routes. The Hudson's Bay Company refused to transport any but its own supplies from the east. Consequently, out of necessity, missionaries such as Lacombe, McDougall and Steinhauer cut their own access roads to wheel supplies to their missions. Typically, once the missions were thriving, the fur trading company opened outposts nearby. Though by 1856 a short northern section was adapted for carts when Fathers Tissot and Maisonneuve had a road cut from Fort Pitt to the Lac La Biche mission, the Lac La Biche Trail was still in the horse/dog-sled stage of development, serving as an access to the buffalo plains to the south and as a supply and market route along the North Saskatchewan for traders and missionaries (see The Carlton Trail).

Missionary affairs had so progressed by the mid-1800s that a network of traders' pack trails and cart roads connected missions across the Alberta-Athabasca districts. (Northern Alberta was a separate district.) Father Lacombe and his faithful Metis guide, Joseph Cardinal, travelled by horseback the 160 miles from Lac Ste. Anne to Lac La Biche. En route they passed Cache and Whitefish Lakes (Rev. Steinhauer's mission), and crossing the Beaver River, they rode along the Lac La Biche trail beside the emerald lake.

Father Lacombe wrote that the weather and scenery were magnificent. The day was cool and even the mosquitoes had disappeared. At the lake they found a large Metis and Cree encampment and were cordially greeted by the natives who had come to trust the clergy in the person of Father Thibault. That night they gathered in a large hut warmed by a fireplace, sang hymns and recited prayers in Cree.

Victoria (founded by Reverend George McDougall in 1863) was perhaps the most important of the crossroad missions, for it was a terminal of transport over the Carlton Trail and for trails leading to the plains to the south, an extension of the trail from Lac La Biche. In 1890, a new route was established from Edmonton to Lac La Biche via St. Albert and the Athabasca Trail (see The Athabasca Landing Trail). The stage was now set for settlers to trek hellish trails to find "Peace."

Fort Assiniboine Trail

Perhaps the story of Alberta's trails should be known as Simpson's Story because the peppery "Little Emperor" blazed so many. The Fort Assiniboine Trail is known as Simpson's Trail for the Hudson's Bay Company Governor George Simpson who ordered it cut in 1823. It was hazardous 80-mile portage linking Fort Edmonton with Fort Assiniboine on the Athabasca River (nearby present-day Whitecourt).

With the merger of the Hudson's Bay Company and the North West Company in 1821, plans were made to extend the trade route through the Great Divide to the Pacific Ocean. The Fort Assiniboine Trail was to be part of this trans-continental trails network. There is a fair amount of controversy concerning the trail's origin. The Fort Assiniboine Trail is believed by some historians to have been Alberta's first deliberately cut pack trail. Still other researchers claim that it was an established trail long before the white traders' arrival. Though perhaps not a primary trail, it served as a secondary Indian portage trail.

Favouring the latter theory is the fact that David Thompson is believed to have used part of the trail in 1799 during a trip from Fort Augustus to Lesser Slave Lake. Leaving Fort Augustus he travelled northwest until he reached the Pembina River which he descended to the Athabasca River. Later, in 1802, he trekked overland to the Forks of the Peace River by the same route. Finally in 1810 when the Peigan Indians threatened to allow no further exploration from Rocky Mountain House, Thompson was forced to blaze a new trail to the Columbia River by way of the Athabasca Pass at the headwaters of the Athabasca River. Though in 1810 Thompson didn't use the Assiniboine Trail, he established the Athabasca River as a link in the trans-mountain route to the Columbia and New Caledonia districts, which in turn led to Simpson's decree in 1823.

Gabriel Franchere also made note of this trail in a journal recounting his journey in 1814 when he descended the Athabasca River from the Athabasca Pass on his way to Montreal with a party of North West Company traders. Franchere mentioned seeing a hunter's lodge located near where Simpson founded Fort Assiniboine ten years later.

Whether the trail existed or not prior to its blazing, Simpson is owed credit for its official development. Simpson chose the route because the Beaver River was almost unnavigable and because he believed the trail to be part of a shorter route via the Saskatchewan River between York Factory on Hudson Bay and Fort George on the Pacific coast. The forthright but vainglorious Simpson fancied his abilities as the ultimate standard of practicality. Simpson was a man of action, notorious for his fast-paced journeys. He believed that the Edmonton–Fort Assiniboine leg of the journey would take a loaded horse only three to four days to complete because it took him only two. In practice, a brigade of some seventy-odd pieces of equipment pulled by two score horses actually needed more than a week to make the trip.

Under the direction of the ambitious John Rowand, later Chief Factor for the entire district, Fort Edmonton began to establish itself as the Hudson's Bay Company's chief post on the prairies. The fort controlled trade in the North Saskatchewan and Athabasca river drainages, and was a consolidation centre for goods moving north to the Peace, west over the mountains and back east to York Factory. There was need, too, for a load-bearing trail from the water terminus at Fort Edmonton to a point on the Athabasca. Fort Assiniboine was founded in 1824, a short distance downstream from the mouth of the Freeman River, 75 miles north-

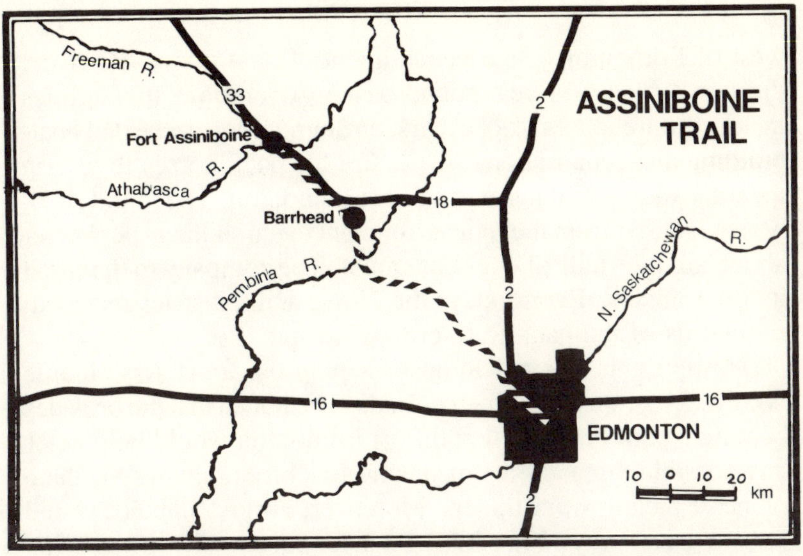
ASSINIBOINE TRAIL
Freeman R.
33
Fort Assiniboine
Athabasca R.
18
Barrhead
2
N. Saskatchewan R.
Pembina R.
16
EDMONTON
10 0 10 20 km

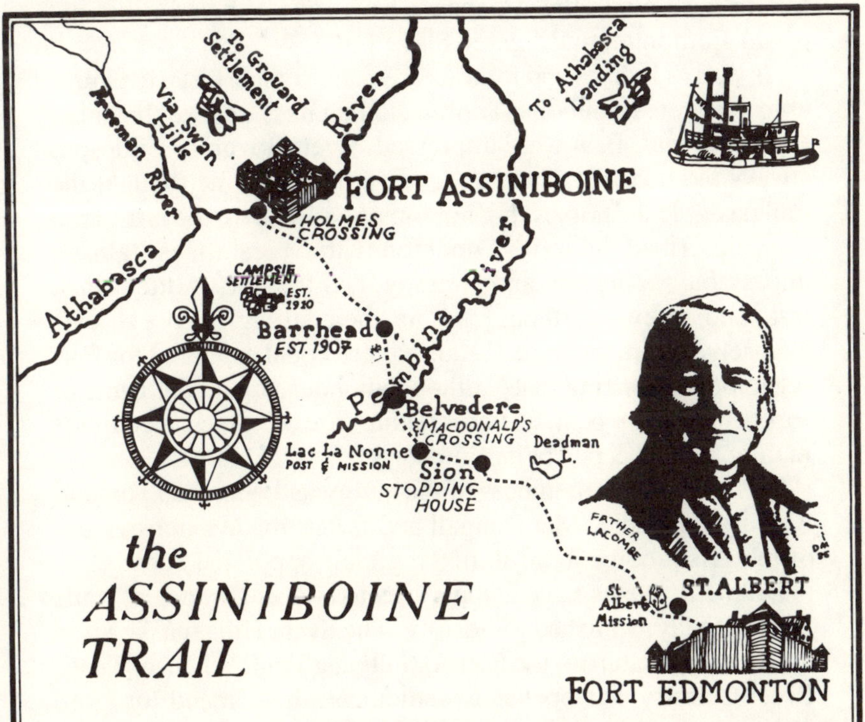
To Grovard Settlement
Via Swan Hills
Freeman River
River
To Athabasca Landing
FORT ASSINIBOINE
HOLMES CROSSING
Athabasca
Pembina River
CAMPSIE SETTLEMENT EST. 1910
Barrhead EST. 1907
Belvedere & MACDONALD'S CROSSING
Lac La Nonne POST & MISSION
Sion STOPPING HOUSE
Deadman L.
FATHER LACOMBE
St. Albert Mission
ST. ALBERT
FORT EDMONTON
the ASSINIBOINE TRAIL

west of Edmonton. On a bench about 30 feet above the river, the fort served as a year-round base of operation. Its facilities included warehouses, root cellars, an ammunition cache and boat-building and repair shops. A pasture for pack horses to use on the trail was cleared just east of Fort Edmonton. The initial stock was purchased from the Indians for eight to ten shillings per horse, about half the value of one beaver pelt. The company then reared its own horses. Even today, the Horse Hills district serves to remind us of the pack train era.

However, it was not Simpson's plan that Fort Assiniboine become a bastion of commerce. Rather, he hoped that the brigades beating a path between it and Fort Edmonton would help deter any possible threat by the menacing Blackfoot, always the bane of the suspicious white traders. Moreover, Fort Assiniboine would serve as a half-way point of dispersion for goods and traffic moving east or west along the Athabasca River or north and south between Edmonton and Lesser Slave Lake.

As well as Indians and fur traders, many notable historic figures used the Fort Assiniboine Trail, including missionaries, botanists and surveyors. Few were impressed. Even Governor Simpson, always the optimist, went as far as to state that he thought the trail passable for horses, but not for sleds. In 1846, the artist Paul Kane described the trail's condition with typical ire as "almost impassable, being wet and swampy, and the horses often stuck fast, and threw off their loads in their struggles to extricate themselves from the mire." Kane voiced equal contempt for Fort Assiniboine, writing that "although honoured with the name of fort [it] is a mere post, used for taking care of horses, a common man or housekeeper being in charge. . . ."

Such illustrious members of the clergy as Lacombe, Thibault Bourassa, Rundle, MacDougall and Evans made extensive use of the Assiniboine Trail in making their appointed rounds.

Dramatic events have always affected the endurance of trails — here today, vanished tomorrow. The Swan Hills muskeg was the dooming feature of the Fort Assiniboine Trail. When the Peace River Country was opened to settlement, the demand for more efficient travel brought about the resurgence of the Athabasca

PROVINCIAL ARCHIVES OF ALBERTA

The Assiniboine Trail was a killer of man and beast. Here a Klondike-bound prospector and packhorse negotiate a crude corduroy road over the muskeg.

Landing Trail, the blazing of the Edson and Hinton trails, and eventually the building of the Edmonton, Dunvegan and British Columbia Railway, the death knell for the Fort Assiniboine Trail.

In the end, the Hudson's Bay Company became frustrated with the deteriorating trail. Chief Factor Richard Hardisty (Rowand's successor) wrote that "the road to Fort Assiniboine is a very bad one, so bad that freighters won't attempt it. Several animals have been entirely ruined going through it in the summer time." Faced with extensive and expensive repairs, the company opted for a new route to Athabasca Landing (see The Athabasca Landing Trail).

Fort Assiniboine had been envisioned by Simpson as an important junction and fort leading into the Peace River Country, and yet in 1859 it was found by a member of the Palliser party to be thoroughly deserted and in total disrepair. Thirty-eight years later during the Klondike Gold Rush, this trail would live in infamy as part of the Edmonton–Yukon "Back Door Route" (see The

Trails To The Klondike).

It should be noted that much of the actual Fort Assiniboine Trail is not easily identified, though the general route is. The season of travel determined its course. Winter travel was easiest over frozen lakes, and the summer path tended to follow local land heights, avoiding natural hazards.

Today, select areas around Alberta proffer a view of the erstwhile route. Departing Edmonton, the Assiniboine Trail passed through the Big Lake Metis settlement (now St. Albert). In settling the area, freeman Metis roamed the northwest ridge overlooking Riviere Qui Barre traversing the height of the land from the North Saskatchewan River to the Athabasca watershed. This area is now the Alexander Indian Reserve. From atop a hill inside the reserve you can see the ridge that the trail followed. The Fort Assiniboine Trail passed just south of present-day Riviere Qui Barre.

Further along at Deadman Lake, the trail took a treacherous turn. Changing water levels would render the trail passable one year, a morass the next. The trail later bypassed the lake along a nearby ridge.

Continuing along the trail, at Scion two stopping houses are still preserved. "Preacher Kelly," the resident minister of a local church and local innkeeper, is buried in the church's cemetery.

Thirty-five miles south of Fort Assiniboine at Lac La Nonne, the fort's fishery, a Hudson's Bay Company post and a Roman Catholic mission were located on the northern shore. The lake was a stopping place from the earliest fur trading days.

Belvedere on the Pembina River was an original ferry site first called MacDonald's Crossing. MacDonald ran a stopping place along the trail, situated on a height of land south of the junction where the Pembina and Paddle rivers meet, running parallel to a modern road allowance. On the north side of the Pembina was the junction of the Fort Assiniboine Trail, the Peavine Trail to the west and a local trail running north and west along the Pembina floodplain in the valley bottom.

Continuing north, the trail passed about a mile east of present-day Barrhead, first settled in 1907, from where it continued through Campsie, a Negro settlement founded in 1910. At the Bethel Baptist Cemetery a cairn honours the community's first settlers.

Onward through the Athabasca valley, the trail followed gently sloping terrain, just a few miles northeast of what is now modern Highway 33.

Holmes Crossing — about seven miles downstream from Fort Assiniboine — was named for William B. Holmes, a Scottish immigrant and ferry operator. A ferry serviced the Horse Creek and Athabasca crossings during the Klondike Gold Rush over the Fort Assiniboine Trail.

The Fort Assiniboine Trail eventually reached Fort Assiniboine, situated on the north bank of the Athabasca River along modern Highway 33.

While today's highway traveller can motor in comfort to Fort Assiniboine, the historic trail lives on in infamy.

Trails to the Klondike

Was the trail from Edmonton a road to riches or a damnable way to hell's wrath? Were the Klondikers who set out from Edmonton attempting a practical route or were they deluded by local boosterism? Does Edmonton deserve its "Gateway to the Klondike" status?

The Klondike chapter of Canada's history is testimony to all who hunger after a dream. The bleak depression of the 1890s had taken its toll. Come hell or high water, the Klondike posed a redemptive chance, and the Klondike Trail. . . well, it was pure adventure, seldom rewarding, usually harsh and too often deadly. One poor devil scratched these words on a sign he hung along the way:

> Hell can't be worse than this trail. I'll chance it.

He later killed himself in despair.

Of the 1500 or so Klondikers who journeyed the Edmonton–Yukon, "Back Door" Trail, many died of starvation or scurvy, scores turned back, a few made it through — only to return with empty pockets — and one gave birth. Six miles out of Fort Assiniboine near the Klondike Trail, a fence surrounds the marked grave site of a baby who died during the gold rush.

There were indeed laws to this rogue trail. Even an ample supply of bannock and beans, warm clothing and hardware, wouldn't suffice if your planning was faulty. Logistics: The prudent man's measure of success. By 1897 when the fabulous rush knew no limits, inspired men everywhere studied their maps. It appeared that there were three possible routes. Those who lived on the west coast (and could afford a ticket) sailed on steamers from coastal ports — Puget Sound, San Francisco, or Seattle — to the coast of Alaska. From there they transferred to light-draft boats on the inland rivers, or continued by means of pack horse, sleigh and foot power through the glacial valleys and snow-crested peaks over Dyea and the Chilkoot Pass.

The Chilkoot Pass was perilous going. It was common practice to hire Indian packers who would help to carry loads and act as guides. Entire Indian families aided with the work, the strongest men shouldering as much as 200 pounds of vital provisions — one poor soul even lugged an organ to the Yukon. The dogs, of course, had their burdens, and when they couldn't negotiate craggy peaks, their masters did the double duty of man and beast.

For inland dwellers, there were two other possible routes: One by water, the other overland. Those who studied their maps found that the straight-line distance from Edmonton to Klondike Creek appeared to be shorter than by way of the coast. As it turned out, distance should have been the least of their concerns, for maps are deceiving.

Still, other qualities favoured Edmonton as the base for the conquest of the Yukon. Edmonton was Canada's most northern community. It lay at the end of the existing network of rail.

For the Klondikers taking the water route, the Landing Trail to the Athabasca River was well-groomed and made for expedient going to the river landing. From here, a fleet of steamboats, scows and anything that would float plied the waterways some 1,800 miles to the mouth of the Peel River down the mighty Mackenzie. Besides, sailing down the Mackenzie River, past its splendid scenery, would be in itself a memorable experience. The hard part was the trip up the Peel, and over the formidable mountain wilderness to Klondike Creek.

The river route had its drawback. It was over 2,500 miles long, a thousand more than what appeared on the map to be the most

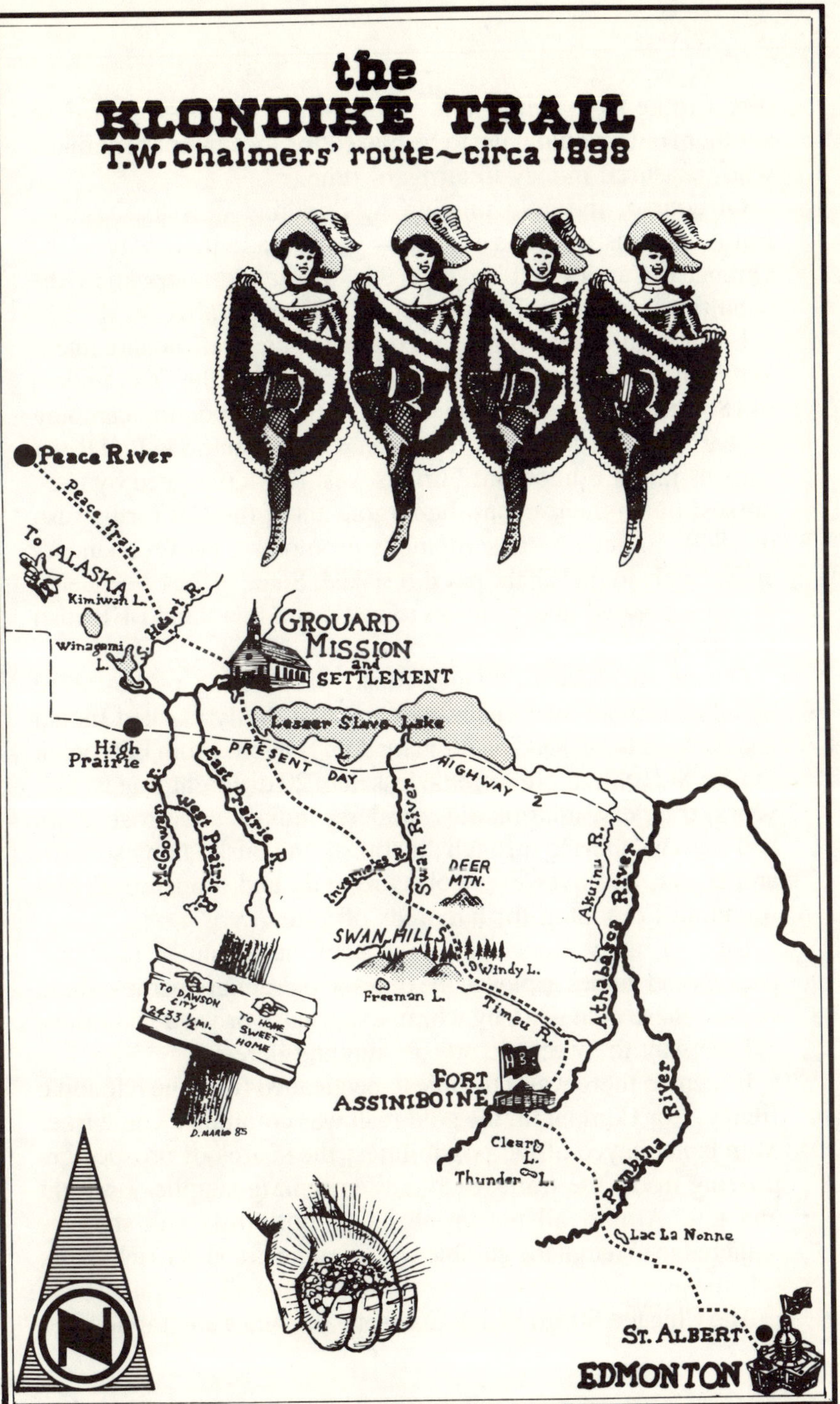

the KLONDIKE TRAIL
T.W. Chalmers' route~circa 1898
Peace River
Peace Trail
To ALASKA
Kimiwan L.
Heart R.
Winagami L.
GROUARD MISSION and SETTLEMENT
Lesser Slave Lake
High Prairie
PRESENT DAY HIGHWAY 2
McGowan Cr.
West Prairie R.
East Prairie R.
Swan River
Inverness R.
DEER MTN.
Akuinu R.
Athabasca River
SWAN HILLS
Windy L.
Freeman L.
Timeu R.
TO DAWSON CITY 2433 1/2 MI.
TO HOME SWEET HOME
H B C
FORT ASSINIBOINE
D. MARIO 85
Clear L.
Thunder L.
Pembina River
Lac La Nonne
ST. ALBERT
EDMONTON
N

direct route from Edmonton. And it was without support. No, another route would have to be blazed for those impatient many who measured money in terms of time.

Of course, the gold mission was nothing new to seasoned Edmontonians who had for many years made their way to the Omineca, Cassier and Yukon fields. These oldtimers knew the conditions one would face to make it through alive.

Early gold rushes through Edmonton are well documented. While making their way to the Cariboo district, the Overlanders of 1862 crossed the prairies just east of Fort Edmonton, camping at nearby Walterdale Flats. They came from Ontario in Red River carts as far as Edmonton. Further west they continued by pack horses, much the same as the Argonauts of the California rush of 1849. Enticed by unscrupulous promoters, the Overlanders arrived only to find all the pay dirt staked. Some of the beleaguered group separated and went on to settle the new land of British Columbia.

During the Klondike Gold Rush, prospecting parties were befuddled by an overwhelming choice of dimly marked Indian and pack trails, all heading off in random fashion. From Edmonton to Fort St. John, there were no less than 20 different routes. For years, trappers, missionaries, traders, Indians and prospectors had been wandering through the forests and bush, along streams and rivers, and over every obscure main and branch trail surrounding Edmonton for hundreds of miles.

Like so many others on this side of the Klondike rainbow, packers and guides exploited the times. Gold-hungry parties would consult these men who, by whim, experience or vague intuition, led the way to sweet victory or stunning defeat.

It seemed there was a fortune to be gleaned from the Klondike frenzy. For Edmonton, the gold rush was commerce come true. Merchants bayed at the possibilities; the scores of prospectors pouring in off the trains each day, snatching supplies up right and left. Amidst all the brouhaha, opportunists and shysters slithered in to tempt the gullible with gimmicks and get-rich-quick schemes.

The Chicago Steam Sleigh Company devised a contraption that

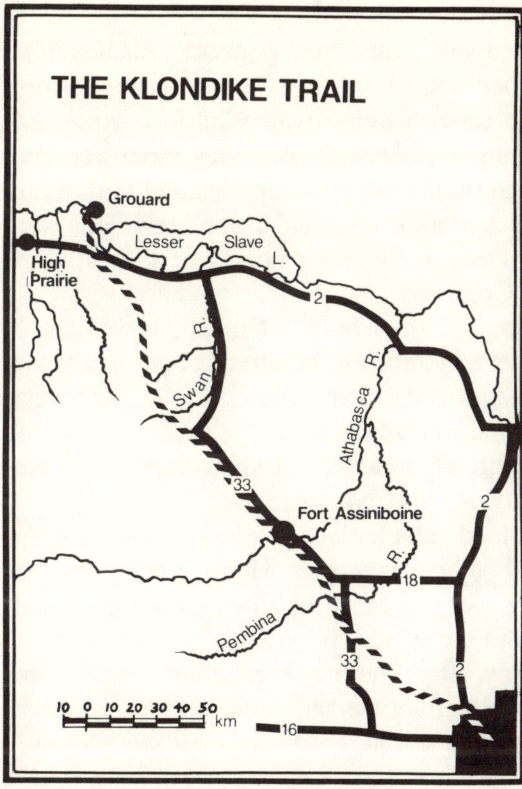

was to revolutionize modern travel. The steam tractor engine had its driving wheels replaced by a steel drum that weighed almost 400 pounds. Projecting from the drum were spikes which were supposed to provide traction. It was intented to pull three cabooses which would not only carry hundreds of pounds of supplies, but also serve as sleeping quarters for a dozen or so men. Its inaugural voyage proved a failure.

Another inventor, Bruno Fabian, tried to patent an amphibious vehicle which, weighing about 500 pounds, was to serve as a boat, sleigh and cart all rolled into one. Its long body was sheeted with galvanized iron and hinged in the middle. It too failed in grand style.

It seemed as though everyone of every condition, great or small, wanted a piece of the Klondike action. Tom Cairns, a local brewer, invented and sold evaporated potatoes. Blacksmiths and craftsmen designed and built every conceivable form of transport designed to carry heavier loads with less expended energy.

As shipping by chartered boats was expensive, most prospectors invested their money in supplies and built their own craft. Conveyances of all shapes and crude conditions were put to the test, though most were of the York boat design used by the Hudson's Bay Company.

Some wise craftsmen built "chameleon" vessels. Not only did each boat have to transport hundreds of pounds of vital supplies, but the most practical crafts were designed to be stripped down and re-formed into small boats for sailing in shallow northern waters. Eventually these were transformed again into sleighs for hinterland trips over mountain passes.

The overland vehicles were no less ludicrously equipped than their water-bound counterparts. Men devised every means possible to roll, pull, carry or bully their hefty loads over such a sizeable distance. But these wee the greenhorns; the experienced prospectors knew that the most reliable means were the least glamorous: sleigh, horse and good old-fashioned foot power — provided of course that there was a path that a sane man might follow.

From Edmonton there arose a cry, "Give us a trail." The Canadian Government, eager to please (perhaps as anxious to stake its own claim) was determined to blaze the best road possible.

The federal government appointed the Royal North West Mounted Police to supervise the blazing. The force, in turn, commissioned Inspector J. D. Moodie. Hope was expressed in Moodie's formal orders that he and his party would reach Klondike Creek by that winter. They did not fulfill their mission until almost a year after the date suggested. By then, the Klondike was old news.

Moodie's project, although a failure, was of two-fold value. Firstly, he successfully explored the possible routes for the first road suitable for wheeled transport to be cut through the Peace River district. Secondly, he kept a detailed account of the journey for reference by future travellers, homesteaders and surveyors

PROVINCIAL ARCHIVES OF ALBERTA

Mrs. Garner was the first woman to pass over The Klondike Trail. She never made it to the Yukon — she and her party were stranded at Spirit River.

to the soon-to-be-opened farming region.

Using pack horses, the Inspector and his party struck out from Edmonton in September of 1897 along the old pack trails via Lac Ste. Anne and Fort Assiniboine. Crossing the Athabasca River near the mouth of the McLeod (Whitecourt), they reached Sturgeon Lake where they crossed the narrows and pressed on to Spirit River in the Grande Prairie region. West of Dunvegan, in the thick forests just west of the Pouce Coupe Prairie, Moodie's guide, a Mr. T. Wilson, lost the trail and then himself. The party searched for three days, following the rambling tracks of a seeming madman, but finally gave up all hope of recovery. Wilson was never heard from again. They kept south of the Peace River until crossing to old Fort St. John November 1. There, Moodie and company killed several of their horses and dried the meat for dog food, replenished the food supply, and re-organized the party. After a month's rest, they again got underway and trekked to the mountains where the trying part of their journey began.

Many Klondike-bound parties took the same doomed trail as Moodie's. The perils they faced — torched prairies, privation and larceny — produced endless tales of hardship. Edmontonians never did favor Moodie's route, but instead wanted a trail cut northwest from the Peace River country to the Nelson River. Later, when the Alaska Highway was laid, it followed the latter route.

The Chalmers' Trail
The "Back Door" Trail From Edmonton

Back at the railhead, the Edmonton Board of Trade was hardly content to rest on its laurels and dispatched its own crew to cut a "superior" trail more to its liking. It was however, the Northwest Territories government that struck paydirt with its choice of T.W. Chalmers to survey a route north. Chalmers' trail was to become the most widely trekked — albeit heartbreaking — trail from Alberta to the Yukon.

An oldtimer by the name of Joe McDonald gave the *Edmonton Bulletin* a harrowing account of life on this trail:

> It was impassable to anything but pack horses, and in order to get through, the parties blockaded there — and they were numerous — had to turn to and cut their own road to Swan River. The trail runs over mountains and hills through country absolutely devoid of feed. . . . For days they (horses) had positively nothing to eat but the back of the trees they were tied to and what little frozen moss they found. As a result the horses dropped in their tracks every day till the trail was landmarked with their carcasses. . .

The more experienced prospectors — those who respected horse flesh — must have cursed the greenhorns when they came upon horses scattered about, dead of exhaustion and ill treatment. Some fools actually had others strap their horses with gear. Fearing they would be unable to remount it, the gear remained until the horses dropped. The going was tortuous for poor dumb beasts, which somewhere down the road shed their bondage once and for all.

Sometimes, though, the "dumb" beasts got the last whinny. Wrote one oldtimer:

> It was a wonderful site to behold. . . horses, pack saddles, flour, pork and beans were scattered all over the hillside. Our horses were not accustomed to pack saddles, and on the long uphill climb our horses laid down, broke their cinch straps and did almost everything possible to get out boxes off.

Yet, despite the horrors on the Klondike Trail, there were good times, wild adventures and enough memories to inspire the tales for many generations to come.

Chalmers was instructed to first explore the Swan Hills region and then report back to Edmonton to organize the cutting party. The trail was to lie between Fort Assiniboine and Lesser Slave Lake. Unfortunately, by the time the road was finished in 1898, the *Edmonton Bulletin* was already referring to the "now historic Klondike rush."

Earlier gold rush parties used poorly marked trails by way of Lac Ste. Anne and Whitecourt. Almost all of the 1898 Klondikers used Chalmers' trail. Before the trail was finished, those who struck out just behind Chalmers used the cut portion, then proceeded ahead on their own, blazing trails as suited their whims.

Chalmers obviously had reservations about the Swan Hills route for in one of his reports he wrote:

> Great fear of their horses being poisoned in these hills (Swan Hills) is felt by many travellers, and they sometimes drive from Salteaux river to Swan river without stopping. The distance is 34 miles and as the road is bad it is most trying on horses. . . This is a mistake, both because of the overstrain on the horses and because being famished when turned out at Swan river they are more apt to eat the poisonous weed, if any is found. . . . If symptoms of poisoning are noticed the horse should be bled. In the mouth (not too far back) is the best place to start the blood, and once the blood is started there is very little danger. If the horses are not famished there is no danger of their eating the weed.

Chalmers' reports of progress in cutting the trail help to identify its course:

> The trail is cut from 10 to 16 feet wide from the Athabasca river to the Lake (Lesser Slave Lake). . . . Goose Lake (Foley Lake), which is about three quarters of a mile off the trail about half way between the ford and leaving of Deep creek, offers splendid feed and a good change to put up hay; but is difficult to find from the trail. From Vermilion creek to Swan river over the Swan hills. . . there is very little feed, except about a mile south of Saulteaux river. Some feed and water may be had at the foot of the hills, a little off the trail. A trail follows Swan river down to the lake shore and then follows the shore of the lake (south shore of Lesser Slave Lake). There is plenty of feed along this route but it is considerably longer than the direct trail, and the creeks, which have to be crossed at their entrance to the lake, are more difficult than where the new trail crosses them.

PROVINCIAL ARCHIVES OF ALBERTA

Gold. It was the irresistible lure of the Klondike. Some struck it rich. These men are escorting $400,000 worth of the precious metal.

Chalmers' Route

By the spring of 1898, engineer Chalmers had cut a road through the Swan Hills almost to Lesser Slave Lake. From Fort Assiniboine the trail veered easterly then north through jackpine until it reached Timeu Creek, near present-day Timeu. From Timeu it struck northwest along Timeu Creek, passing over Deer Mountain and highland points of Swan Hills. (Deer Mountain is about 18 miles northeast of the town of Swan Hills.) Crossing the Swan Hills, the Yukon Road headed northwest over the most mountain-like hills, flanked the northeast slopes of House Mountain until it met the headwaters of the Driftpile River, about 12 miles south of Lesser Slave Lake. After another 25 miles, it met the shoreline of Lesser Slave Lake between what are now Joussard and Enilda. It then flanked the lakeshore to its narrows. Here Klondikers crossed the water to the Lesser Slave Lake settlement on the north shore (now Grouard).

In his article, "Overland Trail to the Klondike," pioneer Frank Walker recounted his journey to the Klondike with a party of six men over Chalmers' trail. Recalling the treacherous hill at Fort Assiniboine, where many prospectors' supplies were lost, he writes:

> At the very top of the hills on a rotten tree had been nailed a piece of board, with a hand pointing in each direction. The one in the direction of the Klondike had underneath: "To Dawson City 2433 1/2 miles" and underneath the other hand said "To Home Sweet Home." Our party of course stood and laughed when they read this and said "This is about right."

Preservation of the Klondike Trail

The Klondike Trail has excellent development potential. Much of the trail, though covered with vegetation, is preserved and deeply etched with markings of the bygone gold rush era.

About 25 miles of the trail — much of which runs through explorative oil leases — from Joussard leading southwest through Swan Hills have been carefully marked with plastic flags. Further south, the trail is covered by brush.

In 1971, a group of interested citizens from the Lesser Slave Lake area organized the Klondike Trail Association with the goal of resurrecting this fine old historical road. A proposal was drafted and presented to various government agencies under whose auspices environmental and recreational concerns are entrusted. Henry Tompkins, past president of the now-defunct Klondike Trail Association, reports that the organization received little in the way of official support or funding for the project. The cost of undertaking such a project on a grand scale is expensive and its benefits to the general public must be duly considered.

Perhaps the draft proposal by the Klondike Trail Association best expressed their cause:

> The historical value of the trail, while recent, is such that it

> should be preserved — showing the hardiness and the vision of people now gone. It is well worth the effort of preservation while there are still a few people left alive who were in their youth connected with the original construction or its usage. At a time when there is increasing public concern with ecology, conservation and the proper use of our environment, the proposal suggests a way to develop the outdoor recreation opportunities in the forested hills without destruction of the natural habitat.

The surrounding forests of the Lesser Slave Lake region are serene and silent, distant from pollution and industrial sprawl. From the perspective of the nearby communities, the natural appeal should be a magnet to draw tourists and local citizens to explore the Klondike Trail in pursuit of outdoor recreation. Such a leisurely purpose would be a radical departure from the frantic, near-desperate fever that originally spawned the trail.

Hell Bent For Peace

"Hell bent for Peace," the battle cry of the early settlers bound for the Peace, applied to the trails, not to the country. The Peace River Country can be justly described as nothing less than majestic. It is a land of wealth, laden with minerals, golden wheat fields, and emerald forests, and fed by the mighty Peace.

The country takes its name from the winding Peace River, so christened by the Woods Cree and Beaver Indians who made peace on its shore in 1782, after a generation of war. As the first western tribe to encounter the white traders from Hudson Bay, the Crees had emerged as effective middlemen, trading goods to other tribes for furs. For many decades, it was not considered necessary for the Hudson's Bay Company to venture inland, as the Crees consolidated the furs and delivered them directly to the bay. As a result, the Crees became a very wealthy and powerful nation, with the advantage of possessing guns while preventing potential competitors from doing the same.

The Crees moved westward either to displace other tribes because they wished to exercise their influence, or because the fur harvest in their traditional lands had depleted the availability of barter goods. Some argue that they were striving to keep a distance from the Europeans. But they did move westward, and for years the battles with other tribes including the Beaver continued, as war parties stole silently along the trails etching the

Peace River drainage. Finally, the Beaver Indians possessed enough guns of their own to withstand the Cree influx, and the two tribes settled their dispute at Peace Point, some 60 miles from the river's mouth.

The Peace River

The land of the Peace River cannot be discussed without devoting a few words to the mighty river itself. The strange traces of dinosaurs in pre-glacial fossils and the visible strata in the valley today reveal the story of a unique geological struggle. The pre-glacial Peace River is the only watercourse to have successfully challenged the barrier of the Rocky Mountains. Elsewhere in North America, the Rockies define the Continental Divide, but

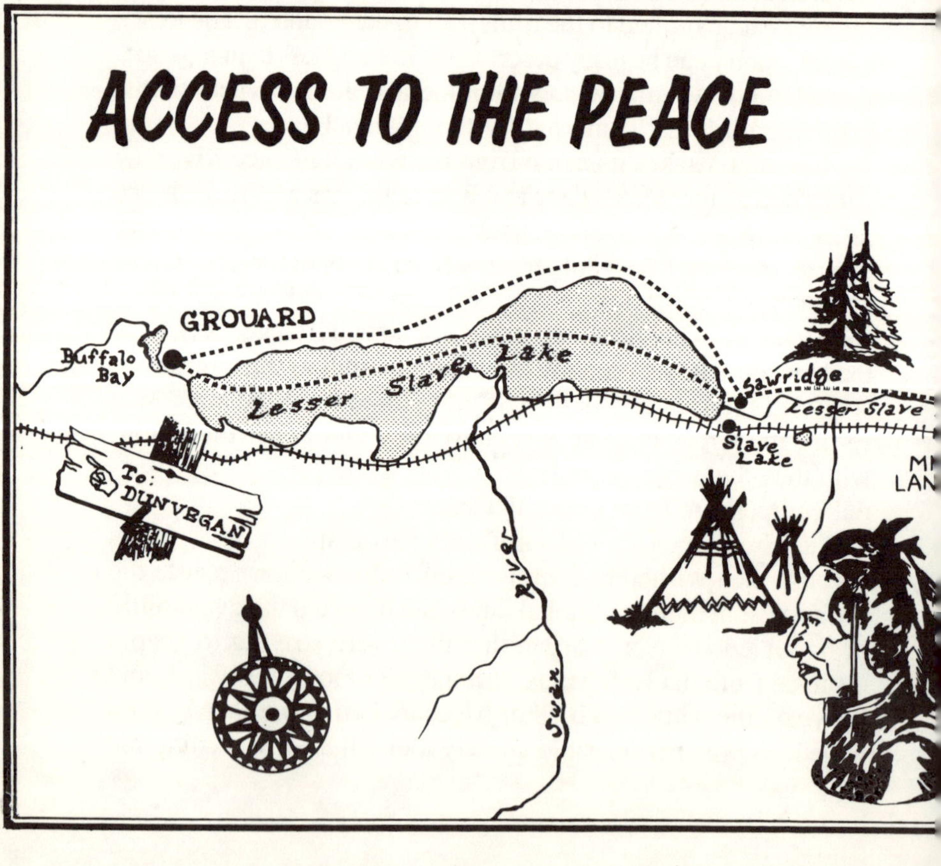

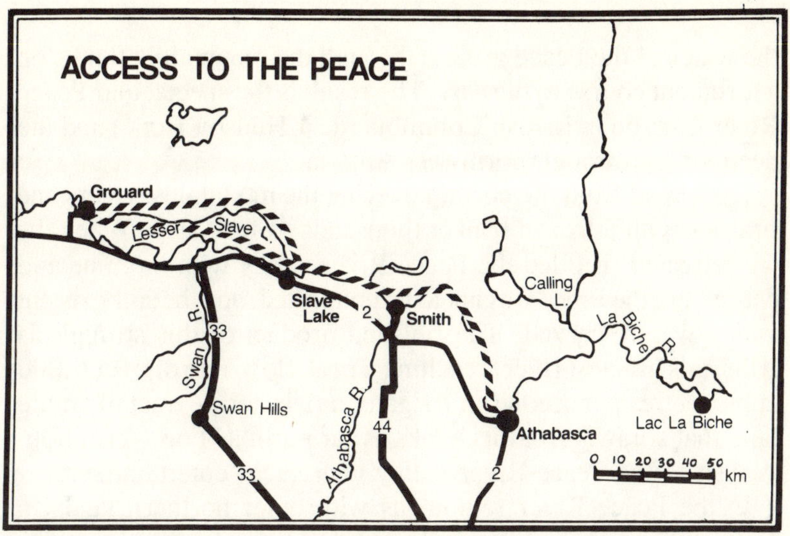
ACCESS TO THE PEACE
Grouard
Lesser Slave L.
Slave Lake
2
Smith
Calling L.
La Biche R.
Lac La Biche
Swan R.
33
Swan Hills
Athabasca R.
44
Athabasca
0 10 20 30 40 50 km

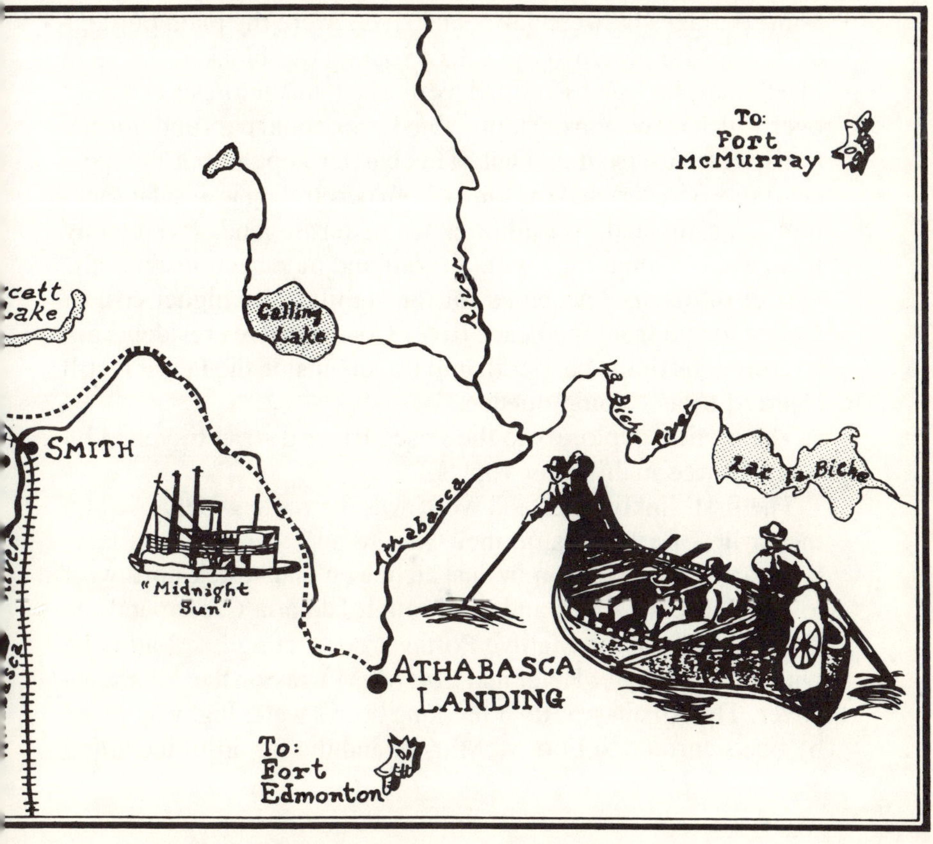
To: Fort McMurray
Calling Lake
River
La Biche River
Lac la Biche
SMITH
"Midnight Sun"
Athabasca
ATHABASCA LANDING
To: Fort Edmonton

the water of the Peace ground through the mountain's flanks on a turbulent course to the east. The result is the spectacular Peace River Canyon in British Columbia (near Hudson Hope) and the deep valley through northwest Alberta.

Ice spread from the north, covering the mountains, rivers and prairies with a frozen banket thousands of feet deep. When the ice retreated, it filled the Peace River valley with moraine and debris. But the Peace was not to be dominated, and the still-existent valley was re-carved. The finished product of this struggle is Alberta's greatest river, reaching a peak flow-rate of over 5,000 cubic metres per second. It irrigates a rich, fertile tract of arable land that sprawls from its banks as far north as Fort Vermilion. A visit to the Peace River valley guarantees entertainment.

In the Peace River region, as with other northern lands in western Canada, water was the chief means of early transport. Short portages between lakes and rivers were the main burdens of travel. More extensive overland routing was blocked by a strip of sub-marginal land scourged by muskeg, uncut forests and mire over which no wagon or stout-legged oxen could pass until formal trails were surveyed and cut. This barrier kept all but the most determined traders and explorers from penetrating its insular sanctum, an estimated 16 million acres of fertile land. Even today in an age of technology with aircraft and modern transport, the barrier of distance accounts for the significantly higher cost of freight to and from the Peace River Country. Area residents are so conscious of the barrier that to travel outside the Peace is still referred to as "going outside."

The earliest explorers to the Peace River district travelled by way of three main water routes.

The first, linking east and west, was the route widely used by the earliest fur traders on their way to and from Hudson Bay. It departed Fort Carlton in Saskatchewan and struck northwest to the Churchill River and on through Isle à la Crosse and the chain of lakes to the Methye Portage, connecting Lac La Loche with the Clearwater River and Fort McMurray on the Athabasca River. The Athabasca River has long been a water highway plyed by boats enroute to Fort McMurray and the far north including

Lake Athabasca and the Mackenzie River, and south to Jasper and over the Athabasca Pass to the basin of the Columbia River. The route was taken by David Thompson in 1810 when he reached the Columbia River, and opened up trade direct from British Columbia.

In addition to being a trail of great historical importance, the Methye Portage — still preserved — is part of a circuit of exciting canoe runs between Lac La Loche and Fort McMurray. All the historical portages along the route have recently been cleared and blazed, and are easily located. The Methye Portage begins at a cairn near a stream which flows into Lac la Loche at Wallis Bay. The trail runs northwesterly, skirting Rendezvous Lake on the west shore, and proceeds to Mackenzie Viewpoint two miles from the Clearwater River. From there the trail descends a steep hill to the river, passing the visible landmark of a firetower across the river, and ends about 13 miles upstream of Whitemud Falls along the Clearwater River.

After reaching Fort McMurray, the route extends down the Athabasca to its delta, and through the boggy Chenal des Quatre Fourches to the mouth of the Peace River. From there the voyageurs faced a 500-mile upstream trip to Dunvegan.

The second historical route to the Peace, by way of the North Saskatchewan River to Edmonton, then overland to the Athabasca, has already been discussed (see The Assiniboine Trail).

The third water route followed the Beaver River through the Cold Lake area to Beaver Lake. From there it was a short portage to Lac La Biche and down the Lac La Biche River to the Athabasca (see The Lac La Biche Trail). From there voyageurs had two possible choices: they could either continue down the river to Fort McMurray, from where they converged with the Methye Portage route, or they could voyage up the Athabasca River to Lesser Slave Lake, cross the lake to Buffalo Bay and trek overland to the Peace River and later the town of its name. Though the Beaver River route was less favoured by traders due to the limits of its freighting capacity, it was advantageous in that it was shorter by hundreds of miles than the Methye Portage. Both led to the upper valleys of either the Athabasca or the Peace Rivers.

Its disadvantages lay with the arduous portages of the upper Beaver River and the long trek to the Peace River from Lesser Slave Lake.

The overland trails were the outgrowth of efforts to expedite travel by sailing the North Saskatchewan River as far west as conveniently possible and then striking out north to the Arctic drainage. The already-mentioned Athabasca Landing and Fort Assiniboine trails were a part of this scheme.

Athabasca Landing to Lesser Slave Lake

Until 1911, the only practical overland route to the Peace was via the Landing Trail to Athabasca, Lesser Slave Lake and cross-country to Dunvegan. In the early days, there were no passable summer roads beyond Athabasca Landing. Only during winters, when the rivers, muskegs and lakes were frozen, was travel eased. Many miles could be covered in a good day's run. In other seasons the trails were exceedingly poor, with little or no grade — just clearings through swamp and muskeg, more suited to amphibious vehicles than to mortal man and horseflesh. Many a horse was disabled by foot rot from plodding in the stew too long.

When making the journey in summer, Athabasca Landing travellers took the steamboat, *Midnight Sun*, to Mirror Landing, across from present-day Smith at the mouth of the Lesser Slave River. Mirror Landing, now vanished, was once a pleasant little hamlet. Near the village, the verdure of forests rose into view. On the waterfront, boats plied the lazy river. Before the Edmonton, Dunvegan and British Columbia Railway left it by the wayside, Mirror Landing was a bustling place, comprised of stores, livery stables, restaurants, inns and the requisite pool halls. Mirror Landing served as an important supply base during construction of the railroad. When the railway passed to the west of the townsite, Mirror Landing slipped into ruin. Except for the ferry which carried Peace River–bound settlers to the north shore, all business ceased.

At Mirror Landing, the Northern Transportation Company, run by James Cornwall (a.k.a. "Peace River" Jim) dispersed freight over a short portage of about 16 miles up the Lesser Slave River from Mirror Landing to Norris or Soto Landing, settled

Fort Dunvegan was a principal supply centre in the Peace River district. Pictured above, a train of horses and wagons hooked in tandem prepares to hit the trail.

by Bill Burns, a woodcutter and ex-pugilist. In those days, the sale of liquor in Athabasca was strictly prohibited, but you could keep an allotment of hooch for personal consumption if issued a license. Bill liked the hair of the dog every now and then, but resented the law's sobering hand, and so was constantly feuding with officers of the peace. One day, Bill and a few boon companions decided to go fishing for a worthy trophy. They set their bait (two empty gin bottles) near a coyote trap in a barrel placed in the bush, and then went off to dip their lines and saw a few logs. A short while later, a constable who had been hot on their trail found their cache. But when he reached in to seize the goods, snap went the trap and his yells rang out. It took some time and a delicate operation for them to find and save the good constable, who all the while cursed in response to their wide-eyed inquiries of "what happened, sir?"

There were many characters not too fondly remembered from that bygone era. "Doc" Wilson was a quack who ran a stagecoach of sorts between Mirror Landing and Bill Burns' place. Doc charged "fureigners" a hard-earned buck and a half to tote their carcasses in his democratic stage the 16 miles over the portage. "All aboard! Ladies first, now. Are you all comfortable? Good!" And off they'd go in grand style. After a quarter of a mile or so they would come to the first of a ridge of steep hills and Doc would regretfully ask that his passengers walk a bit so as not to strain his horses. Begrudgingly they would, but when they mounted the hill, all they would see was a cloud of dust, the stage long gone. Fifteen miles and some sprained ankles later at Soto Landing, they could find no Doctor in the house. He and Bill had no doubt gone fishing.

From Soto Landing, a side-wheeler, the *Northern Light*, plied the waters of Lesser Slave Lake for about 100 miles to Shaw's Point on the lake's west shore. Enroute it either stopped at Sawridge (now the town of Slave Lake) or sailed the open water the entire way to Grouard. There was also a road from Mirror Landing to Sawridge and its trading post, warehouse and customs. These last were in the charge of Tom Lilac, a Justice of the Peace, who, though sworn to uphold the law and quench the flow of contraband whiskey, took many a taste-test before declaring it to be the real McCoy and not just vinegar.

Sawridge might have been modern but it didn't have a ferry, so wayfarers had to swim their horses and stock across the Lesser Slave River. This meant stripping off harnesses and freight and building a raft if none was available. From Sawridge along the lakeshore to the mouth of the Swan River flowing into the the lake from the south shore, there were no passable roads. A trail followed the telegraph line, but the stumps of cleared trees were cut above wagon axle height, so travellers either took the steamer to Grouard or circled the lake's west end in shallow water.

In the winter, travellers headed out straight across the frozen lake to Grouard. Yet for those hell-bent for Peace, the toughest trail was yet to come.

Edson–Grande Prairie Trail

Much has been written of the Edson–Grande Prairie Trail — mostly truth, but some bunkum. Writers, want to spin a good yarn, add a touch of colour to what is otherwise a bald narrative. Suffice it to say that the Edson–Grande Prairie Trail was not the funereal road depicted by writers pie-eyed by the romance of the times. Processions of settlers did not stop to bury dead babies. Grande Prairie settlers did not customarily survive on wild horsemeat. That there was hardship along the trail cannot be denied, but the trail was merely long and difficult, characteristics shared by many others. Still, for the settlers en route to Grande Prairie in 1911, it's a wonder that those "Hell bent for Peace" didn't turn tail, hell-bent for home.

About the year 1908, prospective homesteaders began to muster the courage to assault the 350-mile trip via Athabasca Landing, Lesser Slave Lake and the trail from Grouard to Peace River Crossing (now Peace River). Though some 200 miles further than a crow flies, this was the only practical route to the rich Peace River Valley.

By 1911, Grande Prairie residents demanded a shorter route, and looked to the railroad to support their cause. According to a survey in the 1870s, the railway was to run from Winnipeg to Fort Saskatchewan, 20 miles east of Edmonton. There it would cross the river, pass through St. Albert and Lac Ste. Anne, and cross the Athabasca River at Whitecourt en route to Grande Prairie. Many survey parties had explored the area, studying its topographical and climatic features, but a railway was never built. The unpeopled land did not protest.

However, by 1910, a smattering of diehard pioneers had straggled in to hew their homes from the forests and wrest livelihoods from the virgin soil. Among western Canada's pioneers, those of the Peace River Country were a breed apart. Westward settlement had never trod an easy road, but by comparison, it was organized and convenient. Families moved in file, along beaten trails and open prairies in the south. Other than a handful of explorers, traders and men of God, few souls had ventured so far north as the isolated land of the Peace River Country, where the closest neighbors might be days away.

Fewer yet had settled the Peace with the intention of farming. Trading posts and agricultural missions had for decades thrived around the Peace River barrier, but they were separated from the south by a 200-mile wide stretch of sand and muskeg. At the turn of this century, the Peace River Country remained the only arable Alberta land yet to be extensively settled. The Edson Board of Trade undertook a grand campaign to entice migrant farmers. It proclaimed Edson far and wide as the "Gateway to the Last Great West." The government was willing to stake 160 acres against your ten hard-earned dollars, war veteran's scrip or half-breed scrip that you wouldn't survive three years. But if survival by farming the richest soil in all of Alberta posed few problems, getting there was a trial by fire.

Come what may, Peace-bound folks — single young bucks, land-wise trappers, crusty prospectors and family men with mouths to feed — would somehow endure, but they needed transportation to markets. When surveying parties continued their work across the country, it seemed almost inevitable that either the Grand Trunk or the Canadian National transcontinental railway would cross *la grande prairie*. But it was not to be. Both tracks passed through the Yellowhead beyond Jasper. Of the towns along the lines, the closest to Grande Prairie was Edson, which though only 200 miles to the south, must have seemed light years away.

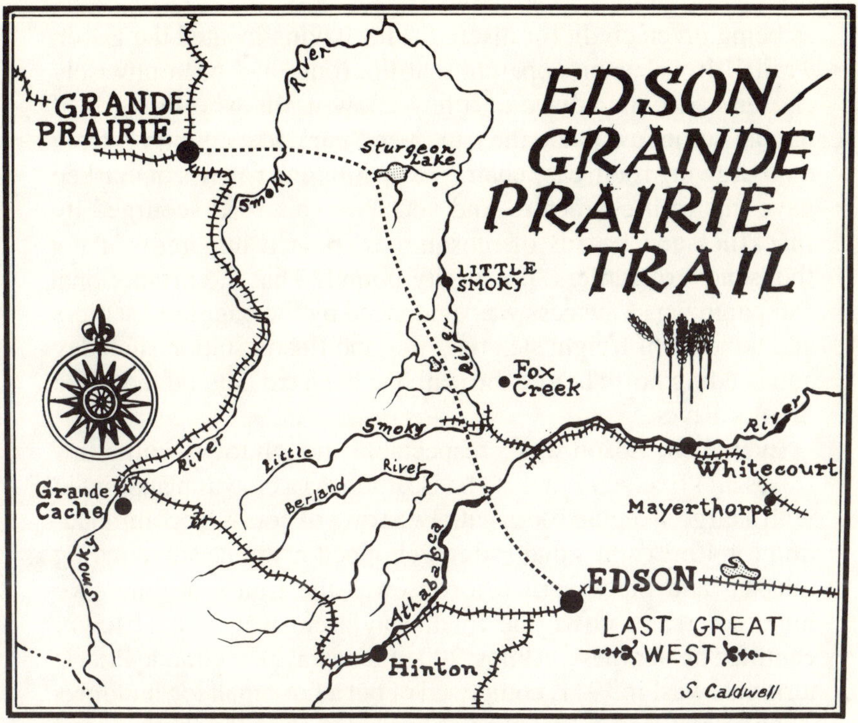

The 200 or so direct miles which separated Edmonton and Grande Prairie were accomplished by the first few settlers who farmed in the summers and spent winters in Edmonton, working in mills or lumber yards. In 1910, when the railroad steamed into Edson, the government had a brainstorm: Why not take some initiative and blaze a trail? Thus was born the much-maligned but famous old Edson–Grande Prairie Trail — no doubt still haunted by cussing spirits stuck in its mud. A. H. McQuarrie was contracted to engineer this marvel, after being cajoled into forsaking his secure position with the Canadian Pacific Railway Bridge Branch and going to work for the Alberta Department of Public Works. The Edson Trail, one of Alberta's first roads to be cut through the bush on a large scale, was to expedite matters until the railway reached the Peace River.

Concerning the project, McQuarrie later wrote: "Nothing in my connection with my work then has ever irked me so much as being given credit (or discredit) for having located the Edson Trail.'' It was soon apparent that the route had been unwisely chosen — by the advice of a few know-it-alls who bragged of having somehow made the trip over ''park-like country'' from their drawing room easy chairs. Steep hills and patches of muskeg gave this trail character. And yet, over this road scourged by blackflies and hoards of mosquitoes, passed hundreds, if not thousands, of settlers, prosperity-bound. That the promotional campaign was a success was evidenced by the gaggle of settlers and boxcars of freight steaming into the Edson station each day and heading north before the trail had been cut beyond the Little Smoky River.

Even then, Edson was a respectable enough town with a few stores and five-star hotels. The Boston was a cozy inn appointed with a large sleeping room with two rows of cots — accommodation for 40. Edson's main street, chopped from the surrounding muskeg and black spruce forest, was the trail's origin. Two hundred strides down the road, wayfarers were out of town, engulfed by darkness. Almost 200 miles away, the Grande Prairie terminus was, in 1911, comprised of but a few small log buildings — one a restaurant, the other a blacksmith shop, but both owned and operated by William Bredin, the only resident.

In the unusually rainy year of 1911, the Edson Trail was a landlocked phenomenon, yet always wet. Harvey Switzer, author of ''On the Edson–Grande Prairie Trail,'' a prize winning article in the Alberta Historical Society's Pioneer Writing Competition, recalled that it rained every day for six weeks solid. Surely Mr. Switzer and party must have thought that even the heavens conspired against them.

When night fell, there was little to do but hang your clothes on a tree and toast them over an open fire. It was not recommended that you choose a spruce tree, as the oozing sap was likely to ignite, sending a tower of fire shooting skyward. It was a sorry sight to see some pantless fool, grovelling for trinkets and change in the ashes. Still, no matter how dreadful the going, humour on the trail was irrepressible. Near the Waskahigan River, a party of Americans carved a face out of a spruce tree, stuck a pipe in

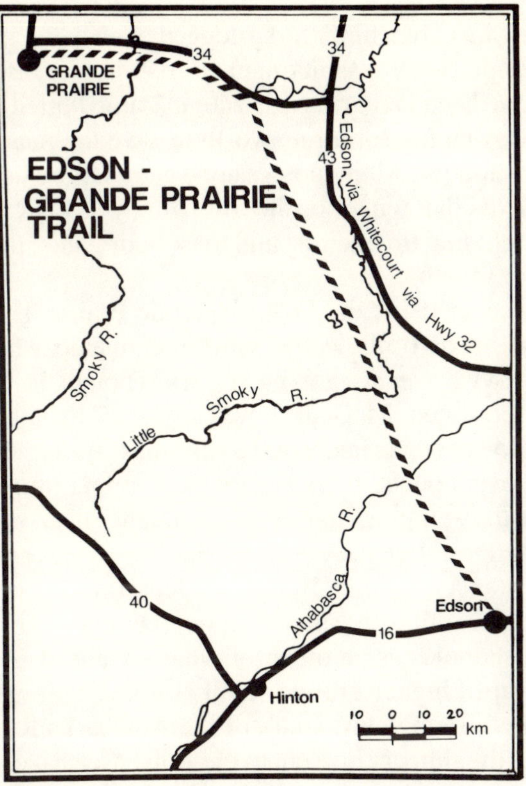

its mouth, transplanted some grass on its head and crowned it with a tattered hat. Then, to lend a touch of indignity, they inscribed beneath the gentleman's gaze: "Uncle Sam, died of mud fever on the Edson Trail 1911."

And then there were hills, always named as befitted their character. "Break Neck Hill," some 40 miles out of Edson where the highland makes its first break down to the benches of the Athabasca valley was the worst of hills along the Edson trail. When tackling icy hills in winter, Grande Prairie–bound folk had to wrap chains around the runners of their sleighs for traction, and chain trees to the rears to prevent the sleighs from overrunning the teams.

Crossing the muskeg bogs necessitated many trips, lugging belongings piece-by-piece, all the while floundering in mud. If the nipping blackflies weren't bad enough, the heat was blistering and the rains drenching. Stout-legged oxen were better suited for such amphibious assaults than skittish horses. But even an ox must have been embarrassed, with its rump butted against the wagon box, digging for all its worth to save the master's piano and earn its fodder. Many possessions were left by the wayside, to be retrieved that winter or the following year. All things considered, including time to rest and toast your pants, the trip took about three months.

Completed in 1911, the Edson–Grande Prairie Trail seemed state-of-the-art as trails went, with its convenient bridges and ferries in service over streams and rivers. Though the project was poorly underwritten, McQuarrie was instructed to build overnight shelters. However, as happened along most trails, enterprising local residents opened their homes as stopping houses.

Most of the stopping houses were situated near water crossings. The furthest stretch lay between the Tony Creek and Sturgeon Lake, a distance of about 50 miles. Departing Edson, the first 25 miles of the trail were smooth-enough going, with only a few scattered mudholes as an omen of what lay ahead. At mile 20, the trail draped higher ground, until at mile 35, it mounted the peak of the hill range just south of the Athabasca River. When negotiating the slippery descent to the Athabasca River (mile 53), wagon teams moved cautiously or tumbled. The Athabasca River was crossed by ferry. From there it was a short jog to the Baptiste (or Berland) River.

From the Baptiste River, the trail ran very straight north-northwest towards the west end of Sturgeon Lake some 70 miles distant. En route it crossed the Little Smoky River and Tony Creek, then passed east of Crooked Lake, angling then slightly west across the Waskahigan River. From Sturgeon Lake, the trail followed an existing route from Grouard to Grande Prairie, ferrying across the Smoky River at Bezanson.

Though a trial for all travellers, the Edson–Grande Prairie Trail carried heavy traffic, especially in winter when the frozen muskeg made easier going (although it meant packing extra supplies). There was even a stagecoach that was put into service by J.B. Taft in 1911; but the traffic had packed down the soft earth so

DAVE JORDAN

The end of The Edson Trail — la grande prairie *as seen today near Sexsmith.*

much that tree stumps made the ride nearly impossible.

The demise of the Edson–Grande Prairie Trail came with the arrival of the Edmonton, Dunvegan and British Columbia Railway at Grande Prairie in 1916.

Grouard–Peace River Trail

The Grouard–Peace River Trail was for generations an Indian warpath and later a supply link between Lesser Slave Lake settlement (now Grouard) and Peace River Crossing (now Peace River). Supplies en route to northwest posts were shipped from Edmonton by way of the Athabasca Landing Trail; loaded on boats plying the Athabasca River to Lesser Slave Lake, its namesake post; then carried further by wagon over the trail to Peace River Crossing and intermediate posts.

The availability of game, fertile soil, water and timber has always determined patterns of exploration and settlement. Of these necessities, the Peace River Country was teeming with more than enough resources to sustain a quality of life. And yet, with the fur trade steeped in a bitter rivalry between the Hudson's Bay and North West companies, "divide and conquer" marketing strategies assumed life and death meaning. Both companies competed for game, each routinely hunting the woodlands and timbered river banks, driving the herds from competitive grounds. Once again, the Peace River Crossing Trail became a warpath of lively skirmishes, with prisoners taken as negotiable pawns in a fur war.

In 1817, the Hudson's Bay Company's Fort Waterloo at Lesser Slave Lake was destroyed during a coup by the North West Company. Always indomitable, the Bay rebuilt near Buffalo Bay at the lake's northwest shore, only 600 feet from their rival's fort.

When the two companies amalgamated in 1821, the Hudson's Bay Company abandoned Fort Waterloo in favor of the new post, dubbed Lesser Slave Lake House (now Grouard).

By the mid 1800s, the Lesser Slave Lake post was the largest settlement beyond Athabasca Landing. It boasted three trading stores — the Hudson's Bay, the Barney Morris, and the Revillon Freres — a hospital, Dominion Land Office, post office, smithy, and boat landing. By 1910, Grouard harboured 447 residents (more than one-tenth of the Peace River Country population), hotels and rooming houses, two restaurants, three poolhalls and stores, the N.W.M.P. district headquarters, a meteorological station and two mission stations. And yet, other than a messenger service that took from 25 to 50 days to reach Edmonton from Peace River Crossing, the Peace had no means of communicating with the outside world.

Among the notable figures who trekked this trail from Lesser Slave Lake to the Peace River was David Thompson. He covered some 50,000 miles by foot and on horse while exploring western Canada, and created a map whose accuracy and detail are admirable even today. According to his preserved journals, Thompson arrived at Lesser Slave Lake on December 16, 1803. The countryside was burnt and smoldering from a recent fire. The trail was choked with fallen timber. He reported an abundance of game along the trail and a large catch of whitefish and pickerel hung to dry at Lesser Slave Lake. Though the soil surrounding Buffalo Bay was poor and attempts at farming unproductive, just a short distance t the north the soil was excellent.

Though the Indians had thrived for generations on the lake's bounty, some white settlers believed that Lesser Slave Lake was drying up and would someday be an empty crevasse, a folly of nature. That the sizeable lake had at one time extended to Grande Prairie seemed to confirm their prophecy. But in light of one poor man's ordeal, the prediction was balderdash. It seems that one merchant, disregarding the advice of others, built his outlet on the shore of the lake. Two years later, customers were paddling to the counter to purchase supplies. From time to time, during dry spells, Buffalo Bay evaporated, leaving a swamp grounded

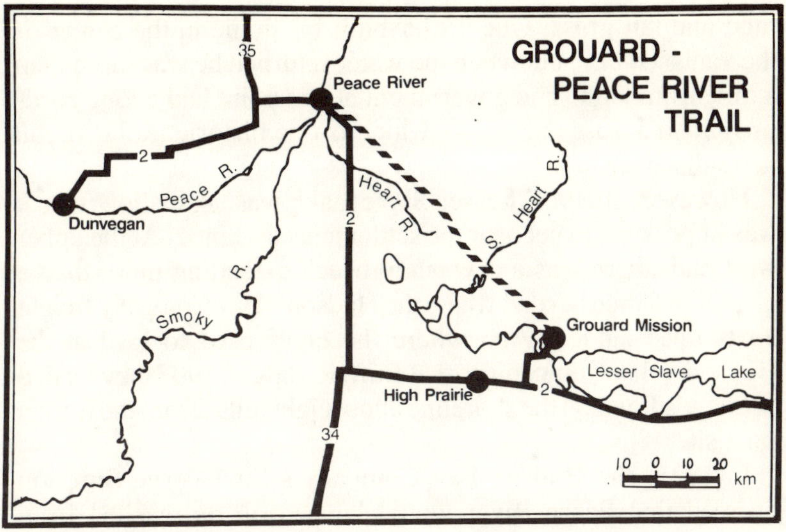
GROUARD -
PEACE RIVER
TRAIL
Peace River
Dunvegan
Peace R.
Heart R.
S. Heart R.
Smoky R.
Grouard Mission
Lesser Slave Lake
High Prairie
35
2
34
10 0 10 20 km

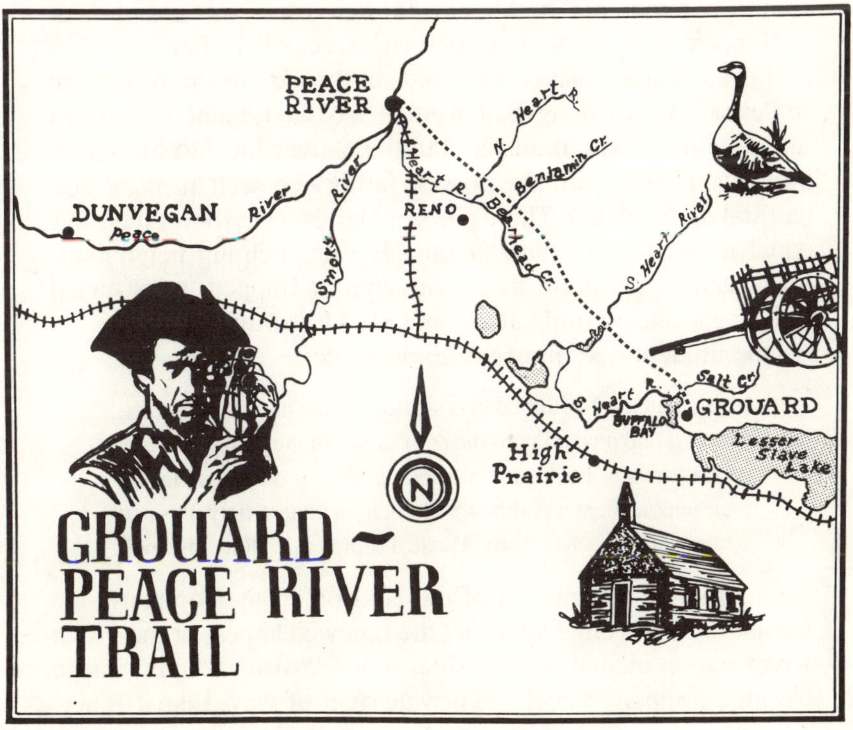
PEACE
RIVER
DUNVEGAN
Peace River
Smoky River
N. Heart R.
Benjamin Cr.
Heart R.
RENO
Bear Head Cr.
S. Heart River
Salt Cr.
S. Heart R.
BUFFALO BAY
GROUARD
Lesser Slave Lake
High Prairie
N
GROUARD ~
PEACE RIVER
TRAIL

mud and tall grass. One settler built his home in the center of the vanished bay, but when the water returned he was more than a mile from land. The government at one point laid a long corduroy road across the bay — which is one historic trail difficult to re-discover.

However, in 1907 Lesser Slave Lake was so shallow that it was impossible to get near the settlements by canoe. At the upper west end, there was a dry sandy beach extending more than a quarter of a mile beyond where the Hudson's Bay Company freight boats once landed. And where the boats used to load at the Hudson's Bay Company post at Buffalo Bay, in 1907 they loaded at Shaw's Point on the shoreline about eight miles from the former post site.

In 1825, the Hudson Bay Company's Sir George Simpson decreed that Peace River would henceforth be reached from Edmonton House, the northwest's gateway and fur trade's bastion of commerce. The Methye Portage route was abandoned.

Henceforth, a stream of missionaries, scientists, homesteaders and even tourists played key roles in the expansion movement to Peace River. Among these were men of God, Fathers Bourassa and Lacombe, who ministered to the natives local to forts Vermilion and Dunvegan. The mission farms were well managed and painstakingly tended. These pioneer clerics were robust men and equal to the task of hard labour. Besides helping neighbours and teaching children, they cultivated or trapped, constructed buildings, blazed trails and generally tamed their parishes.

One unidentified pioneer author wrote:

> No man in the Northland gets so close to the native as the missionary, and no man wins the confidence of the dusky children of the forest as he does. He enters in to their daily life, learns their language, sympathizes with them in their simply joys and sorrows and teaches them practical things as well as spiritual.

Contrary to the interest of the Hudson's Bay Company, the Grouard Trail's chief sponsor, the renewed appeal of the Peace River region incited an intensive, if not destructive, pilgrimage over the company's declared private right-of-way. Like it or not,

The Grouard–Peace River Trail in 1911. The road to Peace River district was a formidable barrier of mud and muskeg.

the trail was open to public use, which meant increased construction and maintenance costs for the trading monopoly.

In 1879, Tom Kerr and 15 labourers were assigned to upgrade the trail into a road suitable for cart passage. This meant re-cutting much of its constricted path, and skirting where possible the treacherous muskegs. Despite such measures, the trail remained a dry-weather road — frequently travelled by cart brigades during the summer months, but otherwise deplorable; for it boasted no bridges, no corduroy roads over the muskeg and no stopping houses. There was only the trail itself, some 90 miles of flat, scorched bushland, plagued by mosquitoes and meat-eating blackflies, torrential rains and unpredictable flash fires.

There was and is, the occasional reward of magnificent scenery from atop the 800-foot high banks of the Peace River valley. From where the Peace River breaks through the Rocky Mountains (which elsewhere in North America defines the Continental Divide), its pre-glacial canyon is a trough-like depression of increasing width and depth. The valley is 60 feet deep at Peace River, averaging in width almost three miles from rim to rim. It widens in spots, leaving small valleys at the bottom, the largest of which lies on the north bank of the river just above "the forks," the junction of the Peace, Smoky and Heart rivers. Here the Peace River makes an abrupt turn to the north and continues so for 300 miles.

In 1886, the Hudson's Bay Company moved its northwest headquarters from Fort Dunvegan to Lesser Slave Lake House. The new location was quite acceptable. Fish were plentiful and the area was an excellent shooting ground teeming with wild ducks and geese. Through the Lesser Slave Lake post filtered all furs taken in the Peace district. Worthy of mention, is the Anglican mission, St. Peter's, founded by Reverend George Holmes, at the northwest end of Buffalo Bay. St. Peter's was the supply depot and central meeting place for all regional Anglican missionaries.

The Hudson's Bay Company kept a large contingent of Red River carts at Lesser Slave Lake House. An expanse of surrounding hay fields fed the oxen. The Red River carts required constant mending. Even they could not withstand the incessant quaking over rock-hard ground, or the sucking grip of muskeg.

By 1893, activity in the Peace River district warranted police attention. A detachment, stationed at Lesser Slave Lake near the Hudson's Bay Company post, routinely patrolled the trail and district, handling disputes and conducting tedious, though "official," business. The Peace was — as befits its name — moderately subdued. Except for the few whiskey traders and sourdoughs brewing trouble, the living was steadfast. Perhaps the biggest scandal to rock the Peace was the Klondikers' unconscionable use of poisoned bait to kill game, a practice which angered the natives, whose roaming dogs (vital to transport in the wintery climate) succumbed in great numbers. The ensuing strife led to the establishment of the N.W.M.P. post at Peace River Crossing in 1898. Regulations prohibiting the sale of liquor and poisoned bait were strictly enforced. In 1899, a treaty was signed with the native people. Alberta Culture–Historic Sites has erected a sign along Highway 2, just east of the turnoff to Grouard, to commemmorate the signing of Treaty Number 8.

In 1900, Jim Cornwall (a.k.a. Peace River Jim) started a regular mail service between Edmonton and Peace River. With stopping houses few and far between, freighters had to carry the feed for their own horses. In 1902, Cornwall obtained a government grant of seven thousand dollars to upgrade the Peace River Crossing Trail. Cornwall and 12 labourers, under the supervision of John Fraser, worked a full summer clearing timber and repairing bridges.

By 1904, a number of stopping houses served travellers and freighters, about half being nearer the Peace River terminus. Most of the owner-operators were trappers, Metis and Indian entrepreneurs. For many years, no white settlers inhabited the land adjacent to the trail. The accommodations consisted of large barns and stables to bed and feed horses, holding as many as six to eight teams. The bunkhouses, more aptly described as people stables, were rustic at best. The horses were pampered while patrons slept on the rough-hewn pole floors and cooked their own meals. Two-thirds of the fee paid to the innkeepers went towards housing the teams. Still, it was always a relief to see the lights of a stopping house around the next turn, where hot meals, comradeship and even a few swigs of contraband whiskey would warm the heart. All drivers also carried grub-boxes which contained food, a frying pan, tea pail and a basin to wash themselves (if so inclined). They dined on the usual fare of flour, beans and smoked bacon (wrapped in paper and covered with tar to preserve freshness between meals).

Freighters charged an average of $25 per person per journey, including freight up to 150 pounds. Baggage was assessed at a further $5.00 per extra 100 pounds.

A short while later, the Red Line Motor Service operated two Ford touring cars along the road. The backseats were removed and replaced with benches installed lengthwise, seating six to eight passengers, baggage at their feet. During winters a bobsleigh, complete with wood-burning stove, covered with canvas and seating ten to twelve passengers, made scheduled runs. In 1913, a 40 horsepower Cadillac covered the distance in just eight hours, compared with the nine days of a few decades before. The Cadillac was driven by A. M. Bezanson and was the first automobile to the Peace.

From Grouard to Peace River, the stopping houses were situated as follows:

At the Heart River settlement (Kapowin) was St. Peter's Mission, the site of which is now marked by cellar holes and two graveyards on a hill overlooking a stand of trees. As noted before, St. Peter's was the supply depot and central meeting place for all regional Anglican missionaries.

At Pakashan, Joe Chalifoux ran a house located about nine miles from Grouard, "Pakashan," meaning "Gambler," was Chalifoux's nickname. The Pakashan Indian Reservation is essentially a family concern, still farmed by descendents of the Chalifoux family.

Joe Lafferty owned and operated a house at Two Creeks.

At the South Heart River crossing, the August Carifelle family ran a house situated approximately one mile east of the South Heart River bridge.

Just south of the South Heart River bridge, there was a small shack used by a telegraph lineman from Grouard.

One-half mile from the telegraph shack, a man named Andrews ran a house.

Spruce Grove, so called because it was near a grove of spruce trees one-half mile west of Spruce Grove Lake, consisted of a house, barn and corrals.

Shell Paul's was situated where the trail entered Little Prairie, about 34 miles from Peace River Crossing.

Capot's Stopping Place, run by Joseph and Nancy Capot, was a horse ranch and may have been in operation as early as the late 1850s.

Sixteen miles from Peace River, Crooked Bridge stopping house — so called either because of a poorly laid section of corduroy road or a nearby bridge built in a half-moon shape — was owned by Peggy Gladue, and comprised of a log house and a couple of barns built during the Klondike Gold Rush.

The last house before Peace River Crossing, just at the base of the Grouard Hills, was known as Four Mile Corner, operated by Benjamin Thomas.

In 1905, the provincial government assumed control of the Peace River Crossing Trail and dispersed funds for improvement — including the building of new bridges, culverts and corduroy roads. Edmonton, Alberta's new capital, was the jumping-off point for a rush of homesteaders to the Peace. The trail, then known as the Peace River Road, was in fair condition, though suitable only for dry weather passage. Approximately $4,000 was spent for construction and maintenance between 1906 and 1908. In his booklet, "The Peace River Trail," published in 1907, author and Peace River pioneer A.M. Bezanson went so far as to state uncategorically that the Peace River "is not a country of dense forests and dismal swamps as some writers of the pseudo-dramatic newspaper articles of recent date would have us believe." Could Mr. Bezanson and party have journeyed the same roads — those rough trails that were the nemesis of so many others. He undoubtedly did, but Bezanson had a vested interest in promoting the Peace River district. He established a townsite in his own name and sold lots from agents coast to coast.

Let us defer to an "objective" report of the *Edmonton Journal*, dated July 17, 1914:

> From Grouard to Peace River landing is the worst trail in the north country at the present time, according to travellers who have come in from that district. The heavy rains have caused the rivers to become so high that many of them have overflowed their banks. . . . On the Grouard-Peace River trail. . . there are several places where the water is from three to five feet deep. At Buffalo Bay, where the country is largely marsh the road is submerged for several miles, the water being three to four feet deep. At Heart River it is worse. The average depth of the water is about four feet, but a number of holes in the trail are so deep that a horse and rider would go completely out of sight. Last week a mounted policeman escaped death. He was travelling at night and attempted to cross the flooded area at Heart. His horse slipped in a hole and fell, pinning him underneath, completely under water. It was only by the assistance of a companion that he was released and escaped with his life.

Route of the Grouard–Peace River Trail

The first organized tour of the Grouard–Peace River Trail was in 1973 by a group of 19 riders as part of the 100th anniversary of the Royal Canadian Mounted Police. Two years later, a group of Lesser Slave Lake locals re-discovered a significant portion of the trail, still preserved in a natural state.

The trail follows Grouard's former main street to a place known as Police Point where it enters a bluff of spruce trees through which it winds for about two miles. Occasional glimpses of Buffalo Bay are granted as the trail hugs the top of an adjacent ridge. There are still remnants of a corduroy road along this section.

The trail then plunges toward Buffalo Bay. A pronounced ditch marks the remains of the trail at the bottom of a steep descent choked with vegetation. As late as 1980, a few scattered poles of the one-time telegraph line also remained. The trail crosses an unnamed creek and veers westward for a mile or so to follow the bank of Salt Creek. (There are no bridges across any of the trail's water crossings.)

From the Salt Creek crossing, the trail's path is again covered with vegetation and wild grass, saplings, rosebushes and thistles. Within the next mile, it crosses three short draws once drained by culverts. It next crosses Hudson's Bay Company land on which a few scattered warehouses stood as late as 1975. The next notable historic site near the trail is that of St. Peter's mission — now vanished and marked by two graveyards on a hill overlooking some cellar holes and a bluff of trees.

A short distance further, it ascends a bench covered by short grass, willows and cranberry. At Wash Creek, near the end of the bench, the remains of a bridge and wagon ruts of one to one-and-a-half feet deep are visible. For a short distance further, the trails is obliterated by plowed fields. It reappears on the Pakashan Indian Reserve where it steadily climbs through poplar and spruce stands.

Here again, deep ruts are preserved. After eight or so miles, near Peavine, it disappears through plowed fields except for a short segment near Peavine Creek.

From Prairie Creek the trail veers northward and disappears under the South Heart River, the area now drowned by a dam. This segment must be crossed by boat. On the other side the trail is fairly clear except where submerged by scattered beaver dams which can cause flooding in portions. Some ruts are visible where the trail is dry. At one section now underwater, the trail makes an abrupt turn to the north. It is confusing if you don't watch carefully.

When approaching Bearhead Creek, the trail passes through a stretch of attractive countryside lined by birch and poplar stands. About five miles past Bearhead Creek, it again crosses cultivated fields for some seven miles southeast of Reno. Further on, much of the trail has vanished except for a few segments which lie on private property.

The trail crosses Benjamin Creek where the remains of bridge pilings, ditches, wagon ruts and a high gradient all mark its passage. The collapsed remains of the Benjamin Creek stopping house may still be visible.

Issues of Trail Preservation

The ideal outdoor recreation plan would integrate rivers and trails. Alberta's waterways serve us in many ways; as transportation routes, for electric power generation, irrigation and for recreation. Our rivers and lakes are inherently worthy of preservation.

Unlike trails, waterways can neither be paved nor plowed under, and yet without a sound policy of conservation in response to the pressures of industrial growth, the beauty of all our natural habitats is unduly threatened.

Under the Corridor Concept, our environment may be more equitably shared by interest groups that vie for resources and recreation. Corridors are tracts of land or water, variable in size, which are managed in such a way as to protect the habitat and ecosystem, as well as geological or historical features within their boundaries.

As applied to waterways, this plan was first developed by the United States. In 1962, the Craighead brothers conducted an environmental impact study which influenced legislation of the Wild and Scenic Rivers Act and consequent protection of waterways for recreation purposes in the United States.

Ontario was the first province in Canada to enact such legislation. Waterways in "Wild River Parks" are now protected from uncontrolled land development. British Columbia has further developed the Corridor Concept in a plan whereby rivers and trails are sheltered by land tracts or buffer zones. Hikers, skiers, bicyclists, and canoeists can enjoy an environment untouched by industrial sprawl.

So far, the only such plans in Alberta have been mere proposals for discussion purposes. These reports do not offer definitive solutions. They are, however, sound overviews which address the issues of trail planning.

In the summer of 1974, Alberta Culture, Youth and Recreation inventoried Alberta's recreational trails. Due to time and money constraints, the project did not include field work, but was instead conducted by mailed surveys and interviews with planning commissions, provincial agencies, local groups and trail guides. Consequently, a thorough catalogue of Alberta's trails could not be developed. The information lacked detail of the condition of trails and their extent of use.

This shortcoming reflects a need mutually voiced by trails enthusiasts. We must upgrade our present knowledge of trail resources if development is to be properly administered. But private groups are reluctant to consult with government agencies unless they are assured that their interests are protected. When these groups request the use of private land for occasional outings, a trust is implied. Abuse of that trust can mean the literal loss of a trail.

In Alberta, the responsiblity of trail management is dispersed among several government and private groups which plan, develop, and manage trails according to their individual priorities and standards. The federal government manages trails in national parks. The provincial government has jurisdiction over trails in provincial parks and forestry recreation areas. Municipalities govern trails in urban and regional parks. Private associations maintain a smattering of local trails.

Of these groups, only governments have the legal authority to police trails against defacement and unsponsored use. Conflicting activities — such as between hikers and cyclists — can lead to hazardous trail conditions. Indiscriminate use can destroy the natural, cultural or historic features of our trails.

Just as old buildings and museum pieces are artifacts of the human past, so too are historic trails, and yet these trails are not acknowledged as heritage resources by Alberta's Heritage Act. Historic trails should be included in the Heritage Act or separate legislation which would provide for the proper identification, safety and protection of such areas.

Unlike museum pieces, historic trails cannot be handily preserved, and unlike recreational trails, historic trails present a special problem in that they cannot be conveniently positioned to appease the interests of surrounding landowners. Historic trails lie where they lie and that we cannot change. The segments of historic trails that we have managed to preserve can be directly attributed to the piecemeal efforts of private citizens who appreciate their inherent value.

Any acts of legislation should provide for land acquisition by outright purchase, leasing of right-of-ways, or through donations. At present, funding for trails is limited. Trails have low priority in government budgets. Private groups pursue funding through special grants, donations and membership fees.

Another pressing question: How much land is realistically needed to protect our trails? Though scenic views may greatly enhance the recreational experience, conflicting interests such as mining, timber or grazing right-of-ways must be duly considered. Private citizens may lend their property to trail groups for occasional outings, but will they tolerate an "invasion" of their privacy on a larger scale? Legislation must guarantee that land rights be protected.

• • •

Lone Pine Publishing welcomes your suggestions for future editions of *Historic Trails Alberta*. Your submissions of maps, photographs, hand-drawn or duplicated sketches detailing existing trails are valuable in helping to unify the effort of historic trails preservation throughout the province. Your support is greatly appreciated and will be duly acknowledged in forthcoming books.

Index